Classic Experiments in Modern Biology

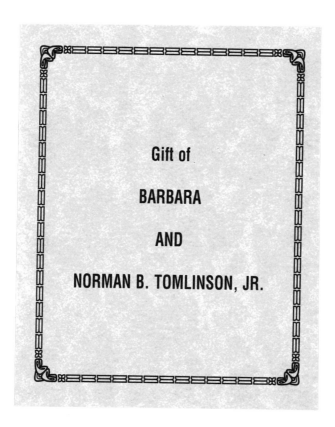

Classic Experiments in Modern Biology

Melvin H. Green

Professor of Biology
University of California, San Diego

W. H. Freeman and Company • New York

Library of Congress Cataloging-in-Publication Data

Green, Melvin H.
    Classic experiments in modern biology / Melvin H. Green.
        p.   cm.
    Includes index.
    ISBN 0-7167-2149-X
    1. Biology—Experiments.   2. Molecular biology—Experiments.
    I. Title.
    QH316.5.G67   1991
    574.8—dc20   90-44304
                                                                        CIP

Printed in the United States of America

The credits for all illustrations appear on pages 193–
196, which constitute an extension of the copyright
page.

1 2 3 4 5 6 7 8 9 0   H   9 9 8 7 6 5 4 3 2 1

Contents

Preface

What is good is doubly so, if it be short; and in like
manner, what is bad is less so if there be little of it.

Baltasar Gracian (1601–1658)

The concept of this book evolved slowly over 15 years of teaching
Biology 1 — The Cell. This course is the first of a three-part series at the
University of California, San Diego (UCSD) that includes Organismal
Biology and Population Biology/Ecology. In Biology 1, the students are
introduced to a vast amount of information from the fields of biochemistry, cell biology, genetics, and molecular biology. As human knowledge
in these fields expanded rapidly, the size of textbooks kept pace.

With this explosion of biological information, my colleagues and I
began to devote less and less time in lectures to the important experiments and technological advances that made this progress possible.
Students are expected to memorize huge amounts of information without gaining much idea about the scientific process that led to this
acquisition of knowledge. All too often, fundamental concepts are totally missed or rapidly forgotten in the race to memorize a myriad of
facts and definitions. As a result, many biology majors are graduating
with little awareness of what it means to be a biologist.

This book is intended to resolve this difficulty in a way that will be
not only helpful, but eye-opening. This relatively short volume de-
scribes many of the most important techniques and examples of "classic" experiments that were conducted from 1958 to 1988 and led to our
current understanding of biology at the cellular and molecular level.
Original photographs, tables, and figures from outstanding research
publications have been carefully selected to illustrate exciting fundamental discoveries. The analysis of these experiments also provides
students with an opportunity to participate in the scientific process of
reviewing data, drawing conclusions, and developing a sense of the

necessity for a critical attitude. Instilling an appreciation for the beauty of simple as well as complex experiments is another major goal of this text. Most important, I hope that it will stimulate thinking about the scientific process in general, not only as a means of gathering information about life, but also as a way of life.

Classic Experiments in Modern Biology was originally intended to be used as a supplement for Biology 1. One hour a week was devoted to a lecture and discussion of experiments in a class of 350 students. This session provided the teaching assistants with an excellent starting point for their weekly discussions with groups of 20 to 30 students. We never had the old problem of facing an audience that could raise no questions, or that asked questions only about definitions of terms. Student response to this course was enthusiastic.

Encouraged by this experience, I next offered a course based solely on this text. It was given a lower division course number to encourage the participation of first- and second-year students, and the only prerequisite was Biology 1. To my surprise, the course was attended by biology majors at all levels, including seniors. We even had one high-school student who wanted an early opportunity to experience a college biology course while taking advanced mathematics on our campus. This course became my most enjoyable teaching experience in over 15 years, and the reviews were unanimously favorable.

Because the experimental orientation of this book does not fit into an obvious niche in the current mode of teaching biology, I offer some further thoughts about its possible use. In my course, the class was divided into groups of five or six students. We met once a week for two hours, covering one chapter in two weeks. Lectures were kept short and served to review basic concepts and techniques covered at the beginning of each chapter. Each week one or two groups presented in detail an experiment described in the text. The students reported in the first person, as though they had been the ones to carry out the work and publish the results. The audience was responsible for raising questions and criticizing the conclusions. The classroom simulated the atmosphere of a small scientific meeting, and every student had an opportunity to participate in the process of science.

Of course, the process of science first and foremost involves the actual doing of experiments from start to finish. This means coming up with a question, designing experiments to answer that question, carrying out those experiments, analyzing the results, and publishing them so that others may confirm or refute them. How much of this is actually done in laboratory courses? I raise this question because I believe that a course based on this book is in many ways better than most lab courses in providing students with a feeling of what a biologist actually does and with an understanding of the scientific method. An innovative use of this text would be as a supplement to a laboratory course that focuses on the molecular and cellular aspects of biology. While getting hands-on experience with fundamental techniques, students could also discuss the classic experiments described here. Biology the course would then be perceived as biology the science.

Having offered instructors some suggestions for using this text, I must now caution students who have had little previous experience in reading scientific papers. Although I have made considerable effort to present the experimental protocols and results clearly and simply, these topics generally are difficult reading. So read slowly, taking time to examine the illustrations in conjunction with the text. You may find that you have to read some sections several times to understand clearly how an experiment was conducted or what the results signify. As with most endeavors, your skill will improve with practice and success will be your reward. So bear with me and think of the words of Albert Einstein: "Everything should be made as simple as possible, but not simpler."

I thank the many students, friends, and colleagues who contributed to the formulation of this manuscript, especially Tammy Tsuchida, Michelle Perello, and Stephen Oppenheimer. Marquerite Richter provided excellent and enthusiastic secretarial support during the many revisions. I also greatly appreciate the financial support offered by the Instructional Improvement Committee of UCSD and the many scientists/educators who supported this endeavor by generously providing original photographs from their research and permission to cite their findings. I extend special thanks to Milton Saier, whose encouragement and enthusiasm for this project continually served as a source of motivation. I am most grateful to my wife, Lynn, and our children, Alex, Arlen, Daniel, and Jessica, who bear the brunt of all my labors during my presence and absence.

Melvin H. Green
December 1990

Classic Experiments in Modern Biology

1

The Power of Microscopy

1.1 Basic Principles of Microscopy

The microscope is perhaps the second oldest instrument used for the study of living systems, the oldest being the knife. Of course, when a knife was used, the system generally didn't live for long. As far back as the early seventeenth century, the great Galileo observed the fascinating compound eyes of an insect with a primitive microscope fashioned from two glass lenses in a cylinder. Although the history of the microscope's development provides very enjoyable reading, it is not our purpose to cover this subject. Nor will we do more than scratch the surface of the way today's microscopes actually function. Rather, this chapter will explore how modern biologists use the microscope to probe the mysteries of life at a molecular level.

One picture may be worth a thousand words, but we have to be able to interpret the picture before we can say anything useful about it. And although seeing may be believing, the interpretation of what is seen through a microscope may vary greatly from one observer to the next. The power of the microscope is greatly enhanced when the various types of microscopy are combined with a variety of biochemical techniques. In this chapter, we will examine some of the classic experiments and methods commonly used to explore the structure of cell membranes and the condensation of DNA (deoxyribonucleic acid) into chromatin and chromosomes. But first let's look at some of the fundamental principles of microscopy and examine some beautiful photomicrographs showing changes that occur in cells during the course of a few diseases.

The units of measurement most commonly used for such tiny things as cells, subcellular organelles, viruses, and molecules are the micrometer (μm), which equals one-millionth (10^{-6}) of a meter; the nanometer (nm), which equals one-billionth (10^{-9}) of a meter; and the angstrom(Å), which equals one ten-billionth (10^{-10}) of a meter. Figure 1.1 shows the sizes of some cells and smaller objects, along with the type of microscope, if any, that is required to view them.

Most bacteria, which are *prokaryotic* cells (cells without a membrane-enclosed nucleus), are 1 to 10 μm in diameter, about 10 times larger than the smallest prokaryotic cells, the mycoplasmas. *Eukaryotic* cells (cells with a membrane-enclosed nucleus) are generally about 10 times larger than bacteria. Eukaryotic cells and the larger bacteria can be seen with the aid of a light microscope. Objects smaller than about 200 nm can be seen clearly only by electron microscopy. These objects include all of the known viruses, *ribosomes* (the subcellular structures on which proteins are synthesized), and even relatively small molecules such as lipids and amino acids.

Basically, an electron microscope works like a light microscope except that the object is visualized by means of electrons rather than light. With a little energy, electrons can be excited to wavelengths of about 0.005 nm. This is considerably shorter than the wavelength of blue light, which is 400 nm. As Table 1.1 indicates, the shorter wavelength gives the electron microscope a considerably greater *resolving power* (RP) than that of the light microscope. The resolving power is inversely proportional to the minimum distance (d) at which two objects can be discerned as being separate (that is, resolved). Imagine two dots that are closer together than the wavelength of light. These dots could not be resolved by a light microscope, but they would appear separated in an electron microscope.

Many clever techniques for the use of both types of microscopes have been devised to study the multitude of fascinating biological structures. A few such methods are shown in Figure 1.2, where a unicellular eukaryote, a paramecium, is the specimen. The cell is shown as viewed with a bright-field, conventional light microscope (Figure 1.2A); a phase-contrast microscope (Figure 1.2B); a scanning electron micro-

Table 1.1. Resolving power (RP).

	Wavelength	Resolving power	Relative resolving power
Human eye	0.4 μm (blue light)	100 μm	1
Light microscope	0.4 μm	0.2 μm	500
Electron microscope (at 50 kilovolts)	0.005 nm	0.5 nm	200,000

RP $= 1/d$ where $d =$ minimum distance at which two objects can be resolved.

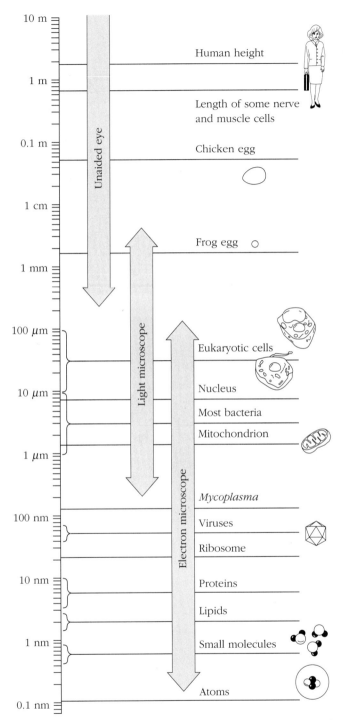

Figure 1.1 The size range of cells, subcellular components, and molecules. Most cells range in size from 1 μm to 100 μm in diameter. Note the sizes of objects that can be observed with the unaided eye, the light microscope, and the electron microscope.

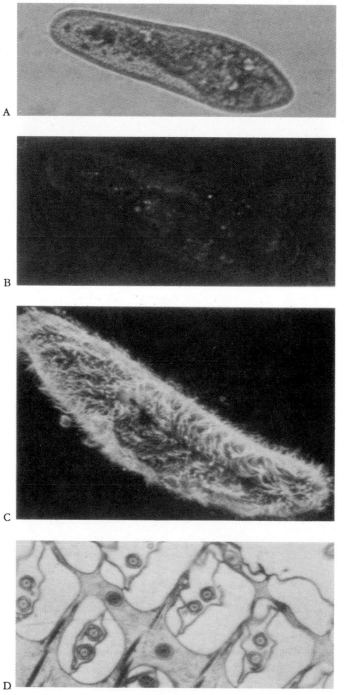

Figure 1.2 Different microscopic images of the paramecium.
(A) Conventional light microscope. (B) Phase-contrast microscope. (C) Scanning electron microscope. (D) Transmission electron microscope. Note the hairlike cilia on the cell surface in (C). When a thin section is prepared for transmission electron microscopy, the cilia are shaved off from the cell, leaving circular "stubs" (D).

scope (EM) (Figure 1.2C); and a transmission electron microscope on a section through the cell surface (Figure 1.2D). The differences in the cell's appearance when it is examined by each of these techniques are quite remarkable.

For the most part, living cells are transparent, being composed primarily of water and tiny subcellular structures. The phase-contrast microscope magnifies small differences in the way that the various cellular structures refract (bend) light, thereby enhancing their contrast beyond what the bright-field microscope can reveal. Membranes, organelles, and nucleoli stand out especially well under phase contrast. Histologists and cytologists have devised many chemical stains for even better visualization of specific structures in cells and tissues.

The scanning EM provides the most dramatic pictures, but is limited to an exploration of a specimen's surfaces. As Figure 1.2C shows, the hairlike projections (cilia) of the paramecium are readily apparent even at the relatively low magnification of 500×. This technique creates a feeling of three dimensions, or depth of field. A beam of electrons is transmitted back and forth across the surface of a specimen that has been coated with a thin layer of metal. The metal emits electrons, which are detected by a type of television camera, and an image (with shadows indicating depth) is formed.

The transmission EM, by contrast, sends its beam of electrons through the specimen. For this reason the specimen must be sliced very thin, so that electron scattering is proportional to the density of the object. The denser the object, the darker it appears in the final image. Note in Figure 1.2D how the slicing of the paramecium has shaved off the cilia, leaving "stubble" (the circular objects). Use of the transmission EM also requires that the specimen be dehydrated, stained with a heavy metal to enhance contrast, and observed in a vacuum. All of these processes can create artifacts that may lead to some heated debates, at the very least, among the scientists who have observed different things in the same structure. Many scientists, for example, once considered the *Golgi complex* to be nothing more than a figment of Golgi's imagination. Thus seeing is not always believing, even if more than a thousand words are used to describe the picture.

1.2 Seeing Signs of Disease

1.2.1 Sickle Cell Anemia

Sickle-cell anemia is one of the best-understood human diseases affecting the red blood cells (RBCs). This genetic disease is found predominantly among people whose ancestry can be traced to Africa, South America, Greece, Turkey, and Italy. Because these areas of the world have been exposed to severe epidemics of malaria, it is thought that the evolutionary persistence of sickle-cell anemia in these populations may be related to the body's attempts to fight off malaria. About 8 percent of black Americans possess a specific mutation in one of the two genes that code for the β-protein chain of hemoglobin. This figure translates into a statistical probability that 1 in 50 children produced by these carriers of the mutant gene, called *hemoglobin S (Hb S)*, will be *homozygous* for the recessive HbS gene; that is, both genes that code for the β protein possess the mutation, and the children inevitably develop sickle-cell anemia. Linus Pauling and his associates discovered that the mutation substitutes a valine (an amino acid) for the glutamic acid (another amino acid) that is normally present in position number 6 of the hemoglobin β chain.

This single amino acid substitution in hemoglobin causes profound clinical effects in the homozygous case. People with sickle-cell anemia exhibit a severe and unrelenting hemolytic anemia that begins within weeks after birth and persists throughout life. Masses of sickled RBCs repeatedly clog vessels in the microcirculatory system, producing considerable pain, progressive organ damage, and often total destruction of the spleen. Children and young adults frequently experience strokes and cerebral hemorrhage. The life span is greatly shortened; most of these people die before the age of 40, after considerable suffering.

The change in hemoglobin structure leads to alterations in the morphology of the RBCs, producing a characteristic sickle shape instead of the usual discoid shape of normal RBCs. Figure 1.3, a scanning electron micrograph of a normal RBC, shows a *reticulocyte* (a precursor to an RBC) being devoured by a *macrophage* (a cell that protects the body from infections and toxins). In Figure 1.4, we can see sickle cells photographed with an interference microscope (Figure 1.4A) and, in the same microscopic field, with a phase-contrast microscope (Figure 1.4B). The variety of shapes that can result from the same mutation is especially striking when the cells are photographed with the scanning electron microscope, as in Figure 1.5. These morphologically altered RBCs wreak havoc on the microcirculatory system of their host.

1.2.2 Cancer

Another disease that alters cellular morphology is cancer. Cancer actually is a group of related diseases that have common biological properties. This disease is thought to originate in a single cell by a genetic change involving one or a very few genes called *oncogenes.* Although the three classes of agents that cause cancer—radiation, chemicals, and viruses—have vastly different natures, each is capable of producing the types of genetic alterations that lead to the conversion of a

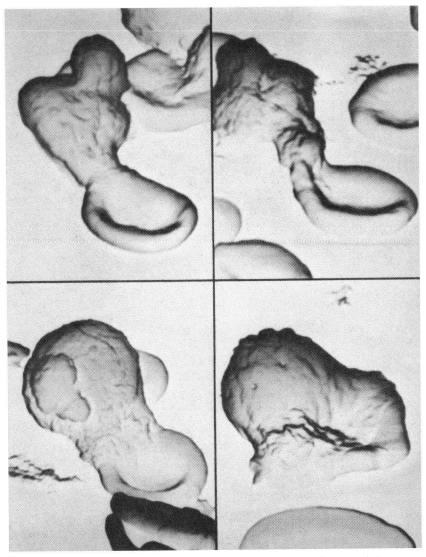

Figure 1.3 Phagocytosis of red blood cells (erythrocytes) by macrophages. Different stages of ingestion are shown.

normal cell into a cancerous (*tumorigenic*) cell. This topic is treated in greater detail in Chapter 4. For the moment we may view a cancer cell as one that has lost the ability to regulate its growth; it divides when it should not and spreads to sites in the body where it does not belong (*metastasis*). Both of these characteristics may result from biochemical changes that affect the properties of the cancer cell membrane. Regardless of the precise mechanism, the result is the development of tumors

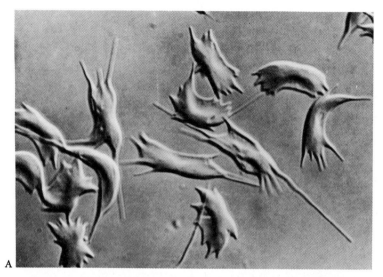

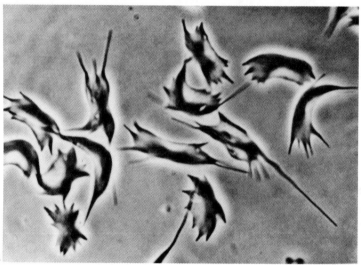

Figure 1.4 Sickle-shaped red blood cells ("sickle cells").
(A) As seen by interference microscopy. (B) As seen by phase-contrast
microscopy.

such as the one seen in Figure 4.14 and the tragic deaths of nearly half
the people who get cancer.

The ability of some viruses to transform a normal cell into a tumorigenic cell resides in a single viral gene. In 1975 Ambros and his colleagues infected chick embryo fibroblasts (undifferentiated cells that
give rise to connective tissue) with a temperature-sensitive (ts) mutant
of a tumor virus called *Rous sarcoma virus* (RSV), which contains ribonucleic acid (RNA). This infection led to the transformation of the

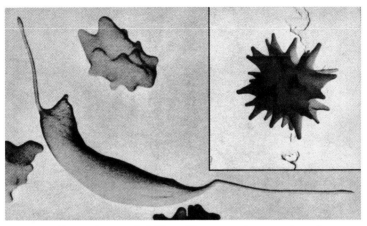

Figure 1.5 Different forms of sickle cells as seen by scanning electron microscopy.
The same mutation in the gene coding for the β-protein of hemoglobin can result in the discocytic (lower left) and echinocytic (right) forms of the red blood cell.

fibroblasts into cancerous cells when they were grown at 36°C. When the cells were grown at the higher temperature of 41°C, however, they exhibited a normal *phenotype* (appearance), having a smooth surface with very few of the protrusions called *microvilli* (Figure 1.6A). At 36°C the phenotype of tumorigenic cells was evident; microvilli and ruffles were very prominent (Figure 1.6B). Thus the inactivation of one viral gene product by heat is apparently sufficient to cause a tumorigenic cell to revert in phenotype to a normal cell.

A second morphological alteration in tumorigenic cells was reported in 1976 by S. J. Singer and his colleagues. These investigators infected a rat kidney cell line with the same ts mutant of RSV as mentioned above. When they directed fluorescent antibodies against *myosin* (a protein that contributes to the formation of filaments involved in the contraction of cells), microfilaments containing this protein appeared as well-organized thick bundles (*stress fibers*) in normal cells (Figure 1.7A), but were disrupted when the cells exhibited the transformed phenotype at the lower temperature of growth (Figure 1.7B). The technique relied on a phase-contrast microscope with filters that permitted detection of the fluorescent antibodies. More on this methodology will appear later in this chapter, in Sections 1.3.3 and 1.3.4.

1.2.3 Infectious Viral Diseases

The electron microscope has literally opened biologists' eyes to the wide world of viruses. Virtually every living animal, plant, and microorganism is susceptible to infection by a wide variety of viruses. The

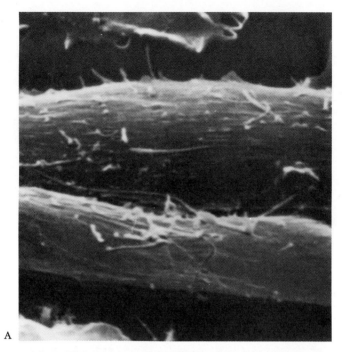

A

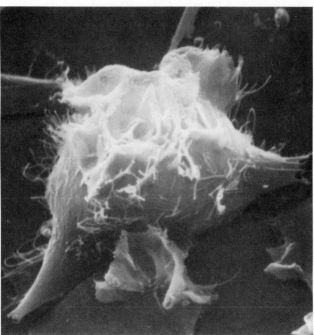

B

Figure 1.6 Normal and transformed cell surface structure in scanning electron micrographs.
Chick embryo fibroblasts were transformed by a temperature-sensitive (ts) mutant of Rous sarcoma virus, which causes the phenotype, normal at 41°C (A), to be transformed at 36°C (B). In the normal state (A), cells have a smooth surface with very few microvilli; in the transformed state (B), both microvilli and ruffles are very prominent. The normal phenotype seen in (A) appeared within two hours after the transformed cells were transferred from 36°C to 41°C.

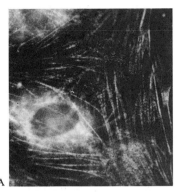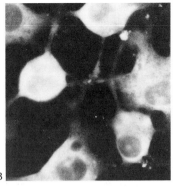

A B

Figure 1.7 Cytoskeleton of normal and transformed fibroblastic rat cells. Normal rat fibroblasts (A) and Rous sarcoma virus transformed derivatives of these cells (B) were stained with fluorescent antibodies that attached to myosin, a protein found in microfilaments. The microfilaments form thick bundles ("stress fibers") in normal cells, but are disrupted in transformed cells. The photographs were taken by phase-contrast light microscopy.

incredible number of viruses, with their diversity of structure and mode of replication, might well qualify the Viridae (viruses) to represent a sixth kingdom in the classification scheme of life forms, which includes Animalia, Plantae, Fungi, Protista, and Monera (prokaryotes). Electron microscopic studies have provided considerable information about the structure of viruses, their mode of entry into the host cell, the way they replicate, the assembly of their molecular components, and their release from the infected cell. From this vast literature I have selected two studies that demonstrate how viruses penetrate and are released from animal cells. These studies tie in with one of our major later topics: the use of microscopy to study the structure of the cell membrane. They also are relevant to two of the deadliest viruses known to humankind: the influenza virus and the human immunodeficiency virus (HIV).

Most of us don't think of influenza, or flu, as a very serious disease, although it certainly occurs frequently and afflicts a large percentage of the population. Except in the aged and in people whose immunity systems are suppressed, flu is rarely fatal in highly developed countries and is often mistaken for the common cold. In addition to symptoms found with a cold, a patient with flu suffers a rapid onset of fever and severe muscle ache. However, worldwide epidemics (*pandemics*) of influenza have been known at least as far back as the early eighteenth century. The outbreak of this disease during World War I developed into one of the worst plagues in history, killing approximately 20 million people throughout the world. Similarly devastating pandemics occurred in 1743 and in 1889–1890. Today much of the lethality of influenza can be prevented with antibiotics such as penicillin and erythromycin. This treatment inhibits subsequent bacterial infections, not the proliferation of the virus itself.

In the first step of infection by any virus, the virus attaches itself to a susceptible host cell and penetrates it. This process is clearly seen by transmission electron microscopy of cells caught at an early stage of infection by the influenza virus (Figure 1.8). Successive stages in the adsorption and penetration of the influenza virus into cells of the chorio-allantoic membrane of a chick embryo are seen in Figures 1.8A–D. Note especially the fusion of the viral membrane envelope with the cell's plasma membrane. These pictures, published in 1968 by Morgan and Rose, strongly support the concept of a fluid membrane (discussed in Section 1.3). Such pictures alone, however, do not rule out the possibility that some other mechanism—say an enzyme that helped create a hole in the plasma membrane—may have permitted the virus to penetrate the cell.

In recent years the human population has been beset by yet another ominous virus: HIV. This virus is highly selective for cells that

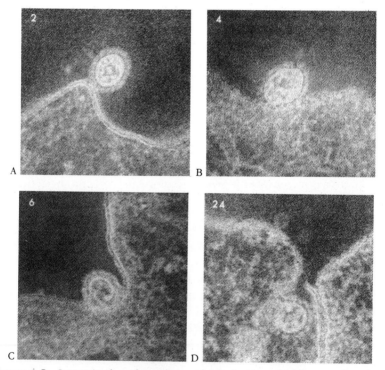

Figure 1.8 Stages in the adsorption and penetration of influenza virus into cells of the chorioallantonic membrane of the chick embryo. Transmission electron micrographs (200,000✕).
(A) Virus is beginning attachment. (B) Viral envelope is fusing with the plasma membrane of the cell. (C) Fusion of the viral envelope with the cell's plasma membrane is more advanced. (D) Viral nucleocapsid has penetrated into the cell's cytoplasm.

participate in the body's immune response, primarily two classes of white blood cells called the CD4 (or T4) lymphocytes and the macrophages. After an extended period of infection, often lasting five years or more, the patient succumbs to one of a variety of other infectious agents because his or her immunologic system has been weakened. Of course, I am referring to AIDS, or acquired (as opposed to inherited) immunodeficiency syndrome. In many people, for reasons not yet known, HIV causes a milder form of this disease known as AIDS-related complex, or ARC. It is generally believed that ARC will eventually develop into AIDS, given sufficient time. Still others infected with HIV are asymptomatic and can be identified only by a blood test that reveals antibodies directed against the AIDS virus. Recent estimates indicate that nearly 100 percent of people who test positive for HIV will eventually develop AIDS and that 50 to 100 percent of those with AIDS will eventually die of this infectious disease or related complications.

In addition to destroying the body's immune system, AIDs commonly causes a variety of neurological disorders that affect both the peripheral and central nervous systems. The severity of this problem is apparent in the fact that over three-fourths of autopsied AIDS subjects are found to have undergone neuropathological changes. The experiments of Dr. Anthony Fauci and his collaborators at the National Institutes of Health (NIH) have shown that mononucleated and multinucleated macrophages support the replication of the AIDS virus in the brain tissue of affected persons (Figure 1.9). The production of viruses by these cells is clearly seen at the plasma membrane, where the viruses are found "budding." This process resembles the reverse of viral penetration. These and other electron microscopic studies have led to the hypothesis that infected macrophages in the brain and elsewhere serve as a reservoir for HIV and as a means of spreading the virus throughout the infected host.

I hope these few samples of microscopic studies in connection with human diseases have whetted your appetite for learning more about the way microscopes can be used to help solve biological problems. Let us now examine in more detail two classic investigations into the structure of cellular membranes and the packing of DNA into chromatin and chromosomes.

1.3   The Fluid Mosaic Model of Membrane Structure

1.3.1   The Meaning of the Model

By now the concept of the fluid mosaic model of the structure of biological membranes is probably very familiar to you. First proposed in 1966 by S. J. Singer and his graduate student G. Nicolson, this model is

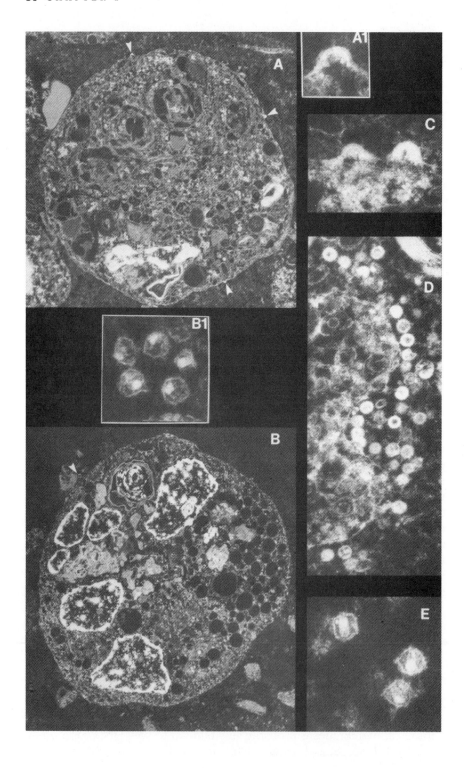

widely accepted as an accurate portrayal of the arrangement of lipids and proteins in most biological membranes (Figure 1.10). The relationship between the phospholipid bilayer of the model and the "railroad track" image seen by electron microscopy of thin sections (Figure 1.11) seems intuitively obvious, if not a proven fact. Presumably the two dark lines that form the railroad track correspond to the electron-dense hydrophilic regions of the 7- to 8-nm phospholipid bilayer. The interior of the track is probably lighter because this hydrophobic region is much less electron dense, so that the electron beams can penetrate more readily.

Before I describe some classic experiments that both led to and supported the fluid mosaic membrane model, we should briefly consider the significance of the terms *fluid* and *mosaic*. The lipid "tails" that extend into the hydrophobic interior of the membrane are long hydrocarbon chains of fatty acids that are linked (via esterification) to glycerol. At biological temperatures, these hydrocarbon chains are in a highly fluid state, bending, rotating, and freely exchanging places with phospholipid neighbors in the same half of the bilayer. The fluidity of the membrane is influenced by the extent of unsaturated fatty acids present, the length of the fatty acid chains, and the amount of another abundant lipid, cholesterol. Detailed quantitative information about membrane fluidity has been obtained from the application of spectroscopic techniques involving the use of fluorescent probes and from electron spin resonance (ESR) and nuclear magnetic resonance (NMR) studies.

The term *mosaic* in the Singer-Nicolson description of membrane structure refers to the protein components. Before this model was developed, the most widely accepted view of membrane structure was that proposed by Davson and Danielli in 1935. Their concept was that of a sandwich consisting of a phospholipid bilayer between two layers of globular, water-soluble proteins. The dark lines of the railroad-track structure seen in electron micrographs such as Figure 1.11 supposedly resulted from the staining of the proteins in the membrane. With the development of more modern techniques for examining membrane

Figure 1.9 Transmission electron micrographs of tissue from a patient with AIDS.
Typical virus-producing mononuclear cells (A: 9300X) and multinucleated cells (B: 5000X), rich in viral particles in varying stages of maturation, are shown. Three particles are budding from the surface of the mononuclear cell (A, arrows); the upper one is enlarged (A1). Two are budding from another cell (C). Late budding and free immature particles with ring-shaped nucleoids are abundant (D). The nucleoids in most mature virions are tangentially or perpendicularly sectioned (B and B1) and only occasional longitudinally sectioned particles showing the conical core shell are seen (E). The outer membrane of the mature virions is usually deformed (B1 and E). (A1, B1, C, and E: 100,000X, D: 50,000X).

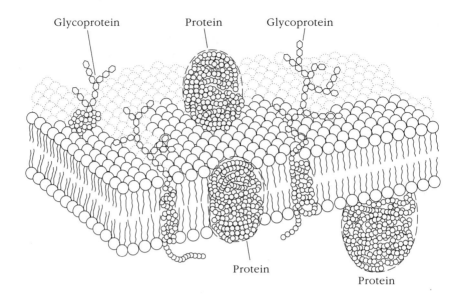

Glycoprotein   Protein   Glycoprotein

Protein

Protein

○| = Phospholipid   § = Amino acid   ∞∞∞ = Carbohydrate
                          chain   unit

Figure 1.10   Fluid mosaic model of membrane structure.
Proteins are distributed throughout a fluid lipid bilayer, with their more
hydrophobic regions embedded in the lipids and their more hydrophilic
regions extending out from the membrane into either the cytoplasm or the
fluid in which the cell is bathed.

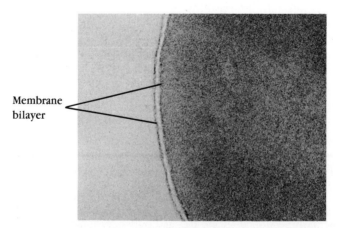

Membrane
bilayer

Figure 1.11   Electron micrograph of an osmium-stained erythrocyte membrane.
Note the "railroad track" appearance of the phospholipid bilayer at the cell
surface.

structure, we now know that a wide variety of proteins are dispersed, with their hydrophobic regions immersed in the lipids and their hydrophilic regions protruding into the aqueous environment on the interior (cytoplasmic) and exterior sides of the membrane (Figure 1.10). These *integral* proteins create a mosaic appearance, rather than forming solid sheets of protein, as the Davson-Danielli model proposed. There are also many *peripheral* proteins that protrude as appendages from the membrane surface. These peripheral proteins are often attached to sugars (*glycosylated*), which increases their solubility in water (*hydrophilicity*).

1.3.2   Freeze-Fracturing and Freeze-Etching

The power of microscopy is greatly enhanced by many techniques used to prepare specimens, such as thin sectioning, staining, and shadowing. The mosaic property of biological membranes was made clear by methods that split open the membrane along the plane of its hydrophobic interior. In this *freeze-fracture* technique, specimens are quickly frozen in liquid nitrogen ($-196°C$). This process minimizes the formation of ice crystals in the tissue and immobilizes cell components. The frozen cells are then fractured with a sharp blow (Figure 1.12). The topography of the membrane surface may be further enhanced by *freeze-etching*, whereby ice that surrounds the cells is removed in a vacuum (*sublimated*). The membrane proteins, which appear as particles in the micrograph, go with one or the other of the lipid layers, leaving indentations in the opposite layer from which the proteins were removed. The figure also illustrates the process of *metal shadowing*, whereby the specimen is coated with thin layers of carbon and then with a heavy metal such as platinum. After the tissue is dissolved with acid, the metal replica, rather than the actual specimen, is used to prepare electron micrographs. The replica is much more stable than the original specimen in the conditions employed for electron microscopy and can be used over and over again.

1.3.3   Seeing Membrane Fluidity

Changes in the shape of animal cell membranes are readily apparent as cells move, extend pseudopods, and ingest material from their environment. While all these morphological changes suggest a certain fluidity of the membrane, the elegant experiments of Frye and Edidin in 1970 provided a clear demonstration of the rapid lateral movement of surface proteins through the lipid bilayer in which they are immersed. These findings created a strong foundation for the fluid mosiac model of membrane structure. These researchers also used several techniques that have widespread applicability in the field of cell biology.

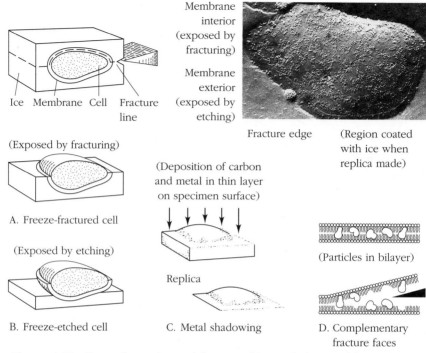

Membrane interior (exposed by fracturing)

Membrane exterior (exposed by etching)

Ice Membrane Cell Fracture line

Fracture edge

(Region coated with ice when replica made)

(Exposed by fracturing)

A. Freeze-fractured cell

(Deposition of carbon and metal in thin layer on specimen surface)

(Exposed by etching)

(Particles in bilayer)

B. Freeze-etched cell

Replica

C. Metal shadowing

D. Complementary fracture faces

Figure 1.12 Freeze-fracturing and freeze-etching techniques with specimens prepared for electron microscopy.
(A) In freeze-fracturing, specimens are rapidly frozen, then fractured by a blow with a sharp blade. (B) In freeze-etching, additional regions of the outer membrane are exposed by evaporation of ice from around the fracture face. (C) Metal shadowing involves coating the specimen with a thin layer of carbon and a heavy metal, generally platinum. The metal coating must be sufficiently thin that details of the exposed surface of the specimen remain visible. The metal replica is used for preparing electron micrographs. (D) The split membrane provides views of the internal structure of both halves of the lipid bilayer. *Inset:* An electron micrograph showing the interior and exterior of a red blood cell membrane prepared by freeze-fracturing and freeze-etching.

Frye and Edidin specifically tagged a surface protein (*antigen*) on two distinct cell types, mouse and human. First they prepared antibodies (Ab) that combined specifically with these surface antigens (Ag), forming Ag-Ab complexes on the cells. To see this reaction, and thereby to localize the surface antigens, they used a technique called *fluorescent antibody tagging*. Rabbit antibodies directed against the human cell surface antigens were obtained from the sera of rabbits that had been injected with human cells. Serum is the fluid that remains after the cells are removed from blood. A preparation of goat antirabbit antibodies (that is, antibodies from goats that react specifically with rabbit Ab) was obtained as a gift from a colleague and was labeled covalently with a fluorescent dye, tetramethylrhodamine (TMR). When the TMR-labeled antirabbit Ab was added to human cells already bound to rabbit Ab, the cells became attached to both types of antibodies wherever the antigenic surface proteins (Ag) were located (see Figure 1.13A). When the human cell was viewed through a microscope with the appropriate filters, the Ag-Ab-Ab-TMR complexes on the membrane fluoresced red. The same methodology was used to localize the surface proteins of the mouse cells, except that the goat antimouse Ab was labeled with an isothiocyanate derivative of fluorescein (FITC). This dye caused the mouse cell membrane to appear green under the appropriate filters (see Figure 1.13A).

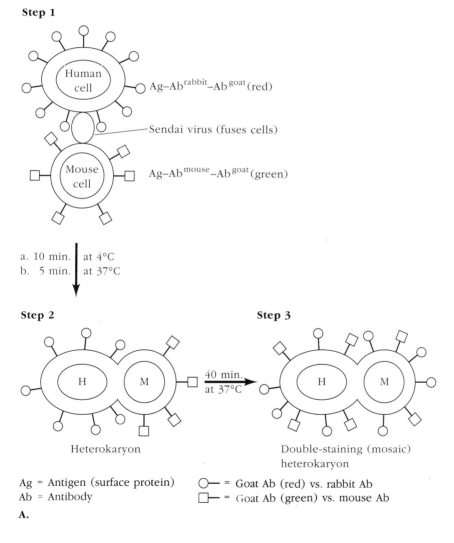

Step 1

Human cell

$Ag–Ab^{rabbit}–Ab^{goat}$ (red)

Sendai virus (fuses cells)

Mouse cell

$Ag–Ab^{mouse}–Ab^{goat}$ (green)

a. 10 min. at 4°C
b. 5 min. at 37°C

Step 2

H M

Heterokaryon

Step 3

40 min. at 37°C

H M

Double-staining (mosaic) heterokaryon

Ag = Antigen (surface protein)
Ab = Antibody

◯— = Goat Ab (red) vs. rabbit Ab
▢— = Goat Ab (green) vs. mouse Ab

A.

Figure 1.13 Experiment of Frye and Edidin: Fusion between human and mouse cells labeled with fluorescent antibodies.
(A) Protocol for experiment. (B) (See following page).

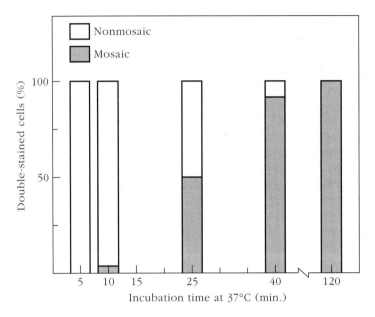

B.

Fig. 1.13 *continued* (B) Appearance of completely double-staining (mosaic) cells in the population of fused cells.

Another important technique was used in these experiments: cell fusion. With the aid of Sendai virus, a membrane-enclosed virus that can attach itself to more than one cell at a time, Frye and Edidin produced cell hybrids by fusing cells from established lines of mouse and human origin. When the mouse and human cells are thus brought into contact, the membranes of the cells fuse, creating a single cell with two different nuclei (*heterokaryon*).

At various times after the cells had fused, the experimenters took photographs of the antibody-tagged cells and examined them to see the position of the red and green fluorescence. Within 40 minutes after cell fusion, the mouse and human surface proteins were completely mixed in nearly all (88 percent) of the heterokaryons (Figures 1.13A and B). This rapid movement of surface antigens slowed markedly when the temperature of the heterokaryons was lowered. Low temperatures are known to reduce the fluidity of lipids and to slow the diffusion of macromolecules such as proteins. Inhibitors of protein synthesis and of ATP (adenosine triphosphate) synthesis did not reduce the rate at which surface antigens migrated. The experimenters thus concluded that "the cell surface . . . is not a rigid structure, but is 'fluid' enough to allow free diffusion of surface antigens" (Frye and Edidin, 1970, p. 319).

In 1981 Sowers and Hackenbrock devised a clever technique to demonstrate that the membranes of organelles, too, have a fluid mosaic structure. A freeze-fracture preparation of a vesicle of the mitochondrial inner membrane revealed particles or aggregates of the integral proteins

(Figure 1.14A). When these same vesicles were placed in a strong electric field, the negatively charged intramembranous protein particles all moved toward one side of each vesicle (Figure 1.14B). The rate of diffusion was similar to that of many proteins in other membranes. When the electric field was shut off, the proteins rapidly returned to their original random distribution.

Quantitative studies on a large number of cell and membrane types indicate that 30 to 90 percent of surface proteins are freely mobile. Surface proteins typically take from 10 to 60 minutes to diffuse through the lipid medium from one pole of a 20-μm-diameter cell to the other. The rate of diffusion in the membranes is from 100 to 10,000 times slower than that of water-soluble proteins in water, because of the greater viscosity of the lipid environment.

1.3.4 The Insulin Receptor and Hormonal Endocytosis

One of the many functions of the plasma membrane is to receive external signals (hormones) that affect cell properties such as growth and differentiation. To underscore the importance of membrane fluidity, we shall consider the phenomenon called *receptor-mediated endocytosis* and the specific case of the well-known hormone insulin. Within seconds insulin stimulates the rate of uptake of glucose into many types of cells. Insulin is a protein that is produced by the B or β cells in the pancreas, and together with glucagon and several other hormones, it is responsible for maintaining the proper level of glucose in the blood.

The first step in the function of peptide hormones such as insulin is to bind to receptor proteins in the cell's plasma membrane. Ira Pastan and his colleagues conducted a detailed study of the binding and internalization of insulin in mouse fibroblasts in 1978. Like Frye and Edidin, they used fluorescence microscopy to follow the movement of proteins in the cell. They prepared fluorescent derivatives of insulin by the covalent addition of either rhodamine alone (*R-insulin*) or lactalbumin labeled with rhodamine (*R-lact-insulin*). The tagged hormone was localized in the mouse cells with the aid of a sensitive image-intensified video camera capable of detecting very low levels of light. Phase and fluorescence micrographs were taken with a Polaroid camera from a television screen that projected an intensified image.

When mouse cells were incubated with the R-lact-insulin at 4°C, the hormone bound in a diffuse pattern all over the cell surface (Figure 1.15). This result was evident in both phase (Figure 1.15A) and fluorescence (Figure 1.15B) micrographs of the same field. These findings suggested that the insulin receptors are spread over the entire cell membrane homogeneously *before* the addition of the tagged insulin, and that the receptors remain in the same place after the hormone bound to them at 4°C. This conclusion was confirmed in a control

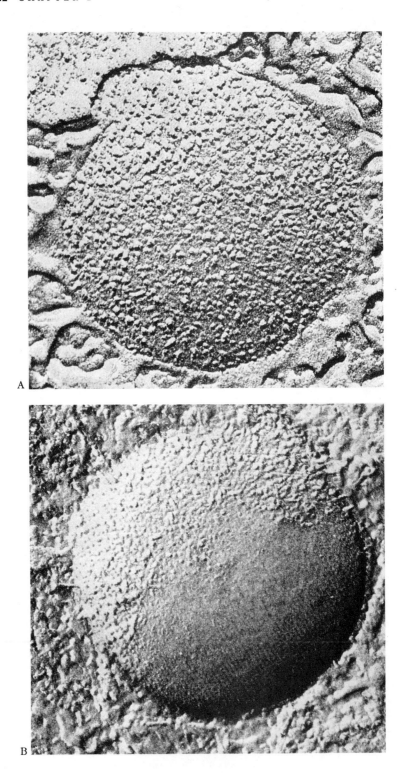

experiment in which the cells had been fixed with 1 percent formaldehyde before the insulin was added. Fixing cells is known to cross-link many proteins covalently, thereby holding them in place.

It was next demonstrated that the occupied receptors for insulin aggregated into "patches" when the cells were transferred from 4°C to 37°C for 15 minutes (Figure 1.16). After this short time, most patches containing the fluorescent R-lact-insulin still appeared to be localized on or near the plasma membrane. This conclusion was supported by the finding that most of the patches could be digested off the cells by the addition of the proteolytic enzyme trypsin.

When the insulin-tagged cells were incubated for longer than 30 minutes at 37°C in a complete growth medium, the hormone molecules were taken into the cells and could be seen in *endocytic vesicles* (Figure 1.17) when they were viewed by phase (Figures 1.17A and C) and by fluorescence (Figures 1.17B and D) microscopy. Note that the fluorescence does not appear over the nuclei, although it did before endocytosis occurred. The hormone apparently enters the cytoplasm but not the nucleus. Pastan and his colleagues postulated that the internalization of insulin first requires aggregation into patches of hormone-receptor complexes, which later become enclosed in membranous endocytic vesicles.

Since they had tagged only the hormone with the fluorescent dye, they could not actually determine whether the receptors for insulin were also present in the insulin aggregates (patches) or whether the receptors were internalized together with the hormone. Nevertheless, the rapid formation of patches is apparently an important prerequisite to insulin action as a signal to the cell. Although these experiments probably raised more questions than they answered, they did bring into focus the relationship between the structural properties of the membrane (that is, its fluidity) and its functional properties (that is, the internalization of a hormonal signal).

1.4   Packing DNA into Chromatin and Chromosomes

1.4.1   The Size and Shape of DNA

Perhaps the greatest technical advance in the field of molecular biology was the technique for viewing DNA molecules by electron microscopy. Early physicochemical estimates of the mass of DNA were

---

Figure 1.14   Electron micrograph of a freeze-fractured vesicle.
Samples were prepared from a mitochondrial inner membrane, before (A) and after (B) the vesicle was subjected to a strong electrical field.

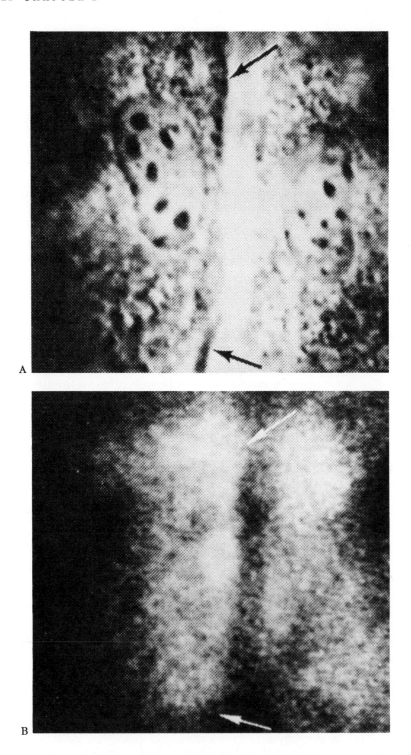

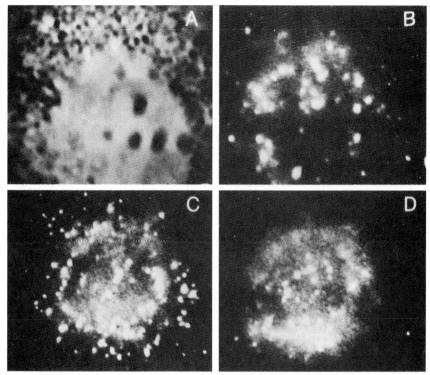

Figure 1.16   Patching of insulin-receptor complexes on cell surfaces. Mouse 3T3 cells were incubated with R-lact-insulin and fixed with formaldehyde. Phase (A) and fluorescence (B) micrographs of the same field (1240X). Fluorescence micrograph of another labeled fibroblast with the objective focused at the bottom (C) and the top (D) of the cell (780X).

on the order of one million daltons (1×10^6 d). We now know that this estimate is closer to the mass of a single gene, and that DNA molecules from prokaryotes such as *Escherichia coli*, or *E. coli* for short, have a molecular weight of about 3×10^9 d. This great discrepancy in mass estimates was due primarily to the fact that the DNA molecule is very shear sensitive. Early methods of DNA isolation involved vigorous stirring and pipetting, which generated large shear forces that broke the molecules into fragments. Once this was understood, it became apparent that electron microscopic techniques would be required to measure the size of the giant DNA molecules that exist in living cells and viruses.

Figure 1.15   Diffuse distribution of insulin bound to mouse 3T3 cells (1240X). Phase (A) and fluorescence (B) micrographs of the same field. Arrows denote portions of outline of cells labeled with R-lact-insulin.

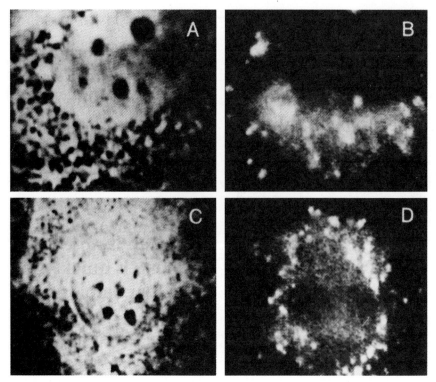

Figure 1.17 Endocytic vesicles containing fluorescent insulin derivatives. Phase (A) and fluorescence (B) micrographs of the same field (1240X) after incubation of cells with R-lact-insulin, followed by 40 minutes of growth at 37°C. Phase (C) and fluorescence (D) micrographs of the same field (620X) after incubation of cells with R-insulin, followed by 45 minutes of growth at 37°C.

The technique for viewing the entire genome of an organism was developed in 1959 by two German microscopists, A. Kleinschmidt and R. K. Zahn. Their dramatic micrograph (Figure 1.18) of a single DNA molecule extruded from the head of a bacterial virus (phage), T2, electrified the scientific world. In this technique, virus particles were disrupted by osmotic shock right on a viewing grid that had been coated with a monolayer of protein (cytochrome c). This protein coating helped the DNA to attach to the grid. No shearing forces were exerted on the DNA, other than the slight ones created by its streaming out of the broken viral head. The DNA was "shadowed" with a heavy metal vapor (platinum) and photographed at a magnification of 80,000. The two ends of the T2 DNA molecule appear at the top center and bottom right of the photograph, and the phage remnants, or "ghost," are evident in the center.

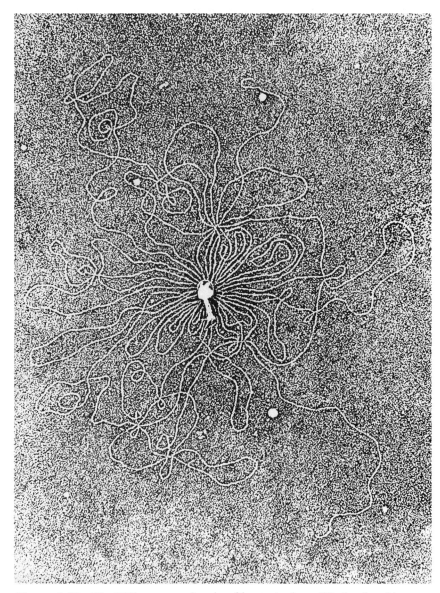

Figure 1.18   The DNA macromolecule of bacteriophage T2 after breaking
open the head of the virus by osmotic shock (80,000×).
Center: The phage "ghost." *Bottom right and top center:* The two ends of
the DNA molecule.

The Kleinschmidt method enabled the molecular weight of DNA to be estimated from the experimentally determined relationship that 1 μm length $= 2 \times 10^6$ d after correction for magnification. From this and other micrographs of T2 DNA, it was concluded that the entire genome of this virus consists of a single linear DNA molecule with a molecular weight of approximately 110×10^6 d. Thus the T2 DNA molecule, with a length of 55 μm and a width of 20 Å (2 nm), must be packaged into an icosahedral head having dimensions of 0.065 μm $\times$ 0.095 μm. (The detailed morphology of the T2 bacteriophage can be seen in Chapter 3, Figure 3.2.)

It was not long before even much larger DNA molecules were observed. In 1963 the renowned microbial geneticist John Cairns demonstrated by a technique called *autoradiography* not only that the *E. coli* genome is a single DNA molecule, but also that it is circular even when it is replicating. Cairns radiolabeled a culture of *E. coli* with tritiated thymidine (^{3}H-TdR) for two generations. Then he gently lysed (broke open) the cells on membranes in a dialysis chamber containing lysozyme, an enzyme purified from egg white. He covered the lysed cells with an emulsion that is sensitive to the β particles (electrons) emitted by the ^{3}H-labeled DNA to form a transparent film. The dark grains that appeared in the film (*autoradiograph*) taken after an exposure time of two months revealed the circular *E. coli* DNA, which had apparently been caught in the act of replication (Figure 1.19).

We shall return soon to the topic of DNA replication. For now let us note that, by this technique, Cairns estimated the length of the *E. coli* DNA molecule to be 1100 μm, which by his calculations was equivalent to a molecular weight of 2.8×10^9 d. Once again we have to wonder how such an enormous molecule can be packaged into a rod-shaped cell measuring only 1 μm $\times$ 2 μm.

The question whether each eukaryotic chromosome has a single molecule of DNA was a logical extension of the findings with *E. coli*. Again electron microscopy combined with autoradiography provided the solution. Bruno Zimm, well known for his physicochemical studies of DNA structure, and Ruth Kavenoff, an expert electron microscopist, chose the fruit fly, *Drosophila melanogaster*, as their system for study. This organism has only four pairs of chromosomes in its diploid set, and, on the basis of DNA content per cell, the largest chromosome would be expected to contain only about 10 times more DNA than that present in one *E. coli* cell.

If indeed there is only one DNA molecule in a *Drosophila* chromosome, it would be far too long to be seen in its entirety in a single electron microscopic field. At the minimum magnification needed to view this DNA, the field is only about 0.01 cm across, whereas the chromosomal DNA molecule has a predicted length of about 1.2 cm. This problem was solved by the technique of autoradiography. The grains created in the film by the emission of β particles from the radiola-

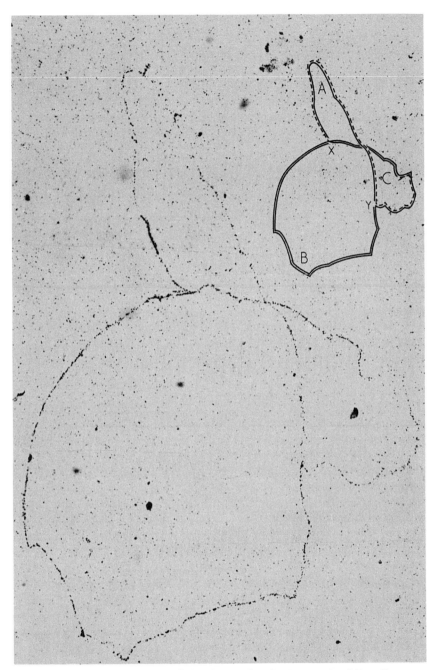

Figure 1.19 Autoradiograph of the DNA molecule (genome) of *E. coli*.
Cells were labeled with tritiated thymidine for two generations and then gently
broken open with the enzyme lysozyme. Exposure time for the autoradiograph
was two months. *Inset:* The same structure is shown diagrammatically and
divided into three sections (A, B, and C) that arise at the two replication forks
(X and Y).

beled DNA are 100 times wider than the actual DNA molecule. Thus it is possible to use a much lower magnification and see the entire molecule in the field of view.

Zimm and Kavenoff labeled cells of *D. melanogaster* with a very small amount of ³H-thymidine and grew them in culture for 24 hours. This "light" labeling of DNA was necessary because radioactive decay can cause breaks in the DNA molecule. The researchers gently lysed the cells with an ionic detergent, taking precautions to avoid shearing the DNA. They avoided enzymatic degradation of DNA by a multitude of methods, including the use of high temperature (50 to 65°C), chelating agents for removing divalent cations needed by DNAases, high salt concentration, proteolytic enzymes, and the ionic detergent. They carefully applied the DNA extract to glass slides and then dipped them in a liquid emulsion sensitive to β particles (electrons). After the slides had been exposed for five months, Zimm and Kavenoff developed them photographically and examined them by dark-field photomicroscopy (Figure 1.20).

These painstaking efforts were amply rewarded. DNA molecules with a contour length of the predicted 1.2 cm were observed. This length corresponded to an estimated molecular weight of between 2.4 and 3.2 $\times$ 10¹⁰ d, the largest DNA ever seen. When this finding was combined with another type of evidence, it could be concluded that, at least for *Drosophila*, "in its simplest form one chromosome contains one

Figure 1.20 Autoradiograph of *D. melanogaster* DNA.
Cultured cells of wild-type *D. melanogaster* were labeled with ³H-thymidine (500 μCi/ml) for 24 hours, washed, diluted 10,000-fold with unlabeled cells, and lysed. Small aliquots were carefully applied to lysozyme-coated slides that were subsequently dipped in Kodak NTB2 liquid emulsion, exposed for five months, developed photographically, and then examined by dark-field photomicroscopy. The contour length, measured with a map measurer, was 1.2 cm.

long molecule of DNA" (Kavenoff, Klotz, and Zimm, 1975, p. 1). The authors carefully pointed out that their conclusion may not apply to all cases, and not even to all *Drosophila* chromosomes, because the existence of giant chromosomes having four times more DNA than the type studied and *polytene chromosomes* having multiple copies of many genes might result in the presence of more than one DNA molecule in a chromosome.

1.4.2 DNA Replication

When we find a molecule as large as DNA so tightly packaged because of the geometric constraints of its environment, we should not be surprised to discover that nature offers a wide variety of solutions to the problem of accurately controlled replication. The discoveries of these solutions probably could have resulted from no method other than direct viewing. Replicating forms of DNA, such as the theta (θ) structures seen by Cairns with *E. coli* (Figure 1.19) and some others that will be described in this section, probably would never have been identified by any other method. The following two examples represent a brief introduction to this topic.

We shall focus first on the nucleolar DNA of the frog *Xenopus laevis* and the work of three Harvard biologists, D. Hourcade, D. Dressler, and J. Wolfson, in 1973. At that time it was well known that the nucleolus is the site of ribosomal RNA (rRNA) synthesis, and that the nucleolar genes that code for rRNA are selectively amplified 1000-fold in the *oocytes* (immature egg cells) that the female frog produces just after she has undergone metamorphosis. The electron microscope was used to investigate the molecular mechanism by which the ribosomal RNA genes, or rDNA, are selectively replicated and thereby amplified in copy number. Electron microscopy was also used to examine the synthesis of ribosomal RNA, as we shall see in Chapter 3.

Amplified rDNA was separated from chromosomal DNA and from mitochondrial DNA by CsCl density gradient centrifugation (see Chapter 3). Using the method of Kleinschmidt and Zahn, with some modification, the Harvard researchers found that the purified DNA contained a small percentage (2 to 5 percent) of circular molecules (Figure 1.21). Measurement of the contour length of more than 100 circles ranging in size from 8×10^6 d to over 140×10^6 d revealed a striking finding. Most of the circles with a molecular weight of less than 40×10^6 d clustered into discrete size classes that were multiples of 8×10^6 d. Other investigators had shown that this is the size of the gene that codes for rRNA.

Furthermore, about one circle in six was a *rolling circle*, such as that shown in Figure 1.22. In this example, the circular portion corresponds in length to three rRNA genes, and the attached tail to 3.4 rRNA genes. In 60 rolling circles that were measured, 85 percent had linear

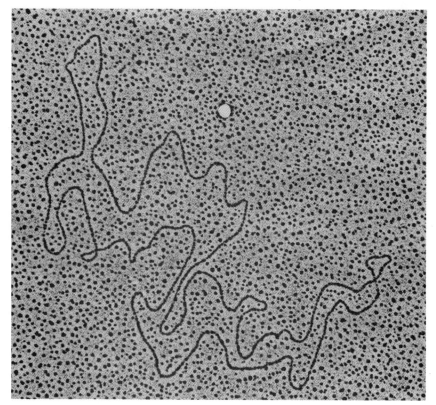

Figure 1.21 A ribosomal DNA circle with a circumference corresponding to four rRNA genes.
The intracellular ribosomal DNA was prepared for electron microscopy by the basic protein film technique of Kleinschmidt and Zahn.

tails that were longer than the rings to which they were attached. It was therefore possible to conclude that these structures did not result from broken *theta structures* (Cairns's replicating forms), which would have either tails that were shorter than the ring or rings that were not necessarily multiples of the length of the rRNA gene.

The authors postulated that the rolling circles are intermediates in the process of ribosomal gene amplification and went on to supply a detailed mechanism. Left unanswered was the question how the first circular rDNA molecule appears in the oocyte. Nevertheless, the appearance of replicating forms of extrachromosomal rDNA was an important discovery that has greatly clarified our concepts of the process of ribosomal gene amplification in oocyte nucleoli. A likely evolutionary role for gene amplification will be taken up in Section 2.5.

Simian virus 40 (SV40) and its cousin polyoma virus are very small tumor viruses that have been used to answer many questions in the fields

Figure 1.22   Ribosomal DNA "rolling circles."
The circumference of the template ring corresponds to 3 ribosomal RNA genes
and the attached tail to 3.4 ribosomal RNA genes. Several single-stranded
phage ϕX174 DNA circles are also visible in this electron micrograph. The ϕX
DNA was included as a standard for length measurements.

of molecular biology and cancer cell biology. These viruses possess an
exceptionally small, closed, circular DNA (3.2×10^6 d), which contains
more than 20 *superhelical* (highly twisted) turns. These twists of the
double helix are introduced before the covalent closure of the two ends
of the DNA. An electron micrograph of a replicating molecule of SV40
DNA, taken at a magnification of 150,000, is shown in Figure 1.23A. It is
evident in this picture and in the interpretive drawing of the DNA
molecule (Figure 1.23B) that there is a superhelical region in the unre-
plicated portion of the DNA molecule. Since a break in even one DNA
strand removes all superhelical turns, SV40 DNA apparently begins replication before the introduction of a break in one or both strands of the
phosphodiester backbone, the repeating units of deoxyribose phosphate
present in DNA.

Electron micrographs have provided considerable information
about the topographic problems that must be solved when a long, double-helical DNA molecule replicates in a confined area such as the
nucleus. As we have seen, SV40 DNA can replicate almost completely

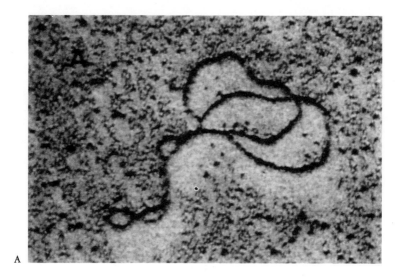

A

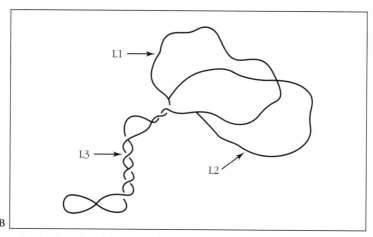

B

Figure 1.23 A twisted SV40 replicating DNA molecule.
(A) Electron micrograph (150,000X). (B) An interpretive drawing of the
molecule. The branches of the replicated portion (L1 and L2) and of the
nonreplicated superhelical section (L3) are indicated.

before it becomes necessary to introduce a break in one of the strands.
The break permits the two circular daughter DNA molecules to separate.
It was also shown that replication often proceeds bidirectionally from a
single starting point. Viruses such as SV40, with their small DNA genomes, have proved to be excellent model systems for learning about
the functions of their more complex host cells.

1.4.3   Chromatin and Chromosome Structure

For many years before 1974, the structural features of cellular chromatin eluded cell biologists because of the notorious difficulty of keeping this material in solution and in a so-called native state. Chromatin was very sticky stuff that readily aggregated and stuck to the walls of containers. It was well known that cellular chromatin contained DNA fibers that were less highly condensed than the DNA present in the chromosomes seen during mitosis and meiosis. It was also known that basic proteins called *histones* were associated with the DNA, at least in regions where the DNA was more highly condensed (*heterochromatin*). The less highly condensed regions (*euchromatin*) were thought to be active in gene expression (that is, RNA synthesis) and were presumed to contain little or no histones.

A relatively slight modification in the preparation of chromatin for electron microscopy led to the exciting discovery of the beads-on-a-string structure. The *nucleosome*, as the bead was called, became visible when the higher-ordered chromatin was diluted with distilled water for a brief time and the sample then fixed with glutaraldehyde. Glutaraldehyde causes a covalent attachment of proteins to DNA and to neighboring proteins, thus serving to fix the existent, presumably native structure. Electron micrographs of two very different types of DNA and their corresponding chromatin structures are seen in Figures 1.24 and 1.25.

In Figure 1.24 we see the SV40 *minichromosome* photographed in its native (A), beaded (B), and deproteinized (C) states. In this research, performed in 1974, J. D. Griffith purified SV40 chromatin from infected monkey cells by a procedure developed for polyoma virus by M. H. Green in 1970. It was quite exciting and surprising to discover that the chromatin structure of viruses is very similar to that of their host cells. The nucleosomes in SV40 chromatin, each 110 Å (11 nm) in diameter, were joined by DNA bridges roughly 20 Å in diameter and 130 Å long. From these measurements Griffith concluded that the SV40 genome was condensed sevenfold by the formation of nucleosomes, each of which contained 170 base pairs of DNA and was separated from the next nucleosome by about 40 base pairs. The solubility of SV40 chromatin, presumably attributable to its relatively minute size, greatly facilitated this research.

Figure 1.25 shows three stages in the packaging of DNA in micrographs of extrachromosomal rDNA molecules (ribosomal genes) purified from oocytes of the water beetle *Dytiscus*. In all three stages — the free DNA molecule (A), the beaded chromatin (B), and the more highly condensed 200 Å fiber (C) — the DNA is the same length. As in the case of SV40 chromatin, we have here another example of chromatin that is very active in the process of RNA synthesis in the cell. Such studies led to the realization that histones serve as structural units of organization in chromatin and helped to dispel the widely held notion that histones function as repressors of gene expression.

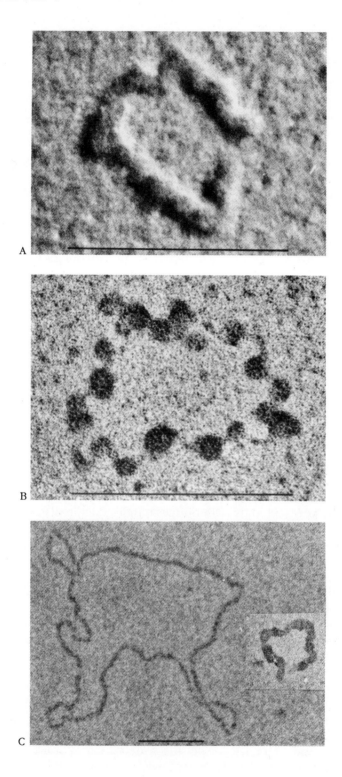

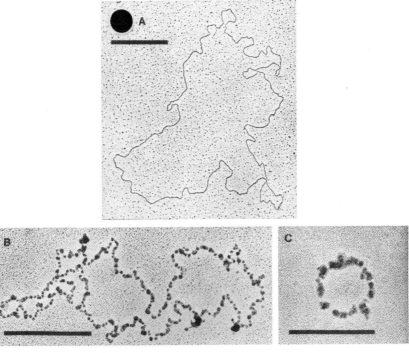

Figure 1.25   Three stages in the packaging of DNA, illustrated by extrachromosomal rDNA molecules from the oocyte of the water beetle, *Dytiscus*.
(A) Free DNA molecule. (B) DNA and histones in a beaded nucleosome condition. (C) Supercoiled state or "200 Å fiber." In each case the length of DNA is the same. Bar, 1 μm (A); 0.5 μm (B and C).

Later studies have led to the following model for the progressive stages of DNA condensation into metaphase chromosomes (Figure 1.26). It should be noted that "naked" DNA (Figure 1.26A) probably does not exist as such in a eukaryotic cell. Fully extended chromatin (11 nm in diameter), with its characteristic beads-on-a-string structure at low ionic strength (Figure 1.26B), can be converted into a more compact fiber, 30 nm in diameter (Figure 1.26C), by an increase in ionic strength or by the addition of divalent cations. This fiber is thought to result when the extended chromatin is coiled in a way that creates six nucleosomes per turn of the double-helical DNA. Further coiling of this fiber results in *looped domains* with a diameter of 130 nm (Figure 1.26D). This is the

Figure 1.24   SV40 minichromosomes and DNA.
High-resolution electron micrographs of the SV40 minichromosome in its native (A), beaded (B), and deproteinized (C) states. The inset in (C) compares a native minichromosome at an equal magnification with the deproteinized DNA.

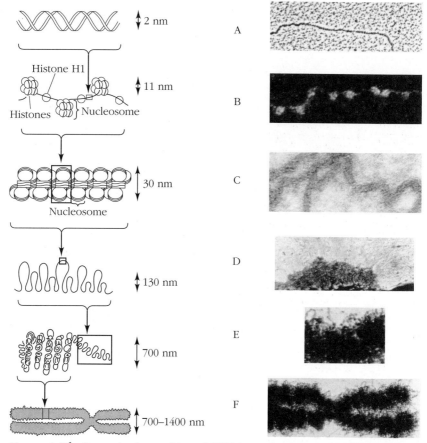

Figure 1.26 Stages in the packing of DNA into chromatin and chromosomes. (A) Double-helical DNA. (B) DNA winds around basic protein particles to form nucleosomes. Each nucleosome is composed of two molecules each of four different types of histone. A fifth type of histone is generally found associated with the DNA between the nucleosomes. (C) The 30-nm chromatin fiber appears to be a tightly wound coil having six nucleosomes per turn. (D) The 130-nm chromatin fiber consists of looped domains of the 30-nm fiber. (E) Further condensation results in a 700-nm structure. The most highly condensed regions, called heterochromatin, are present during interphase and are virtually inactive with regard to the production of RNA. (F) The metaphase chromosome, 700–1400 nm in diameter, still displays chromatin fibers. At this stage of mitosis, virtually no RNA is produced by the DNA.

structure that predominates in nuclei during *interphase*; that is, all stages of the cell cycle other than mitosis. Also evident throughout interphase is the still more highly condensed form called heterochromatin (700 nm in diameter) composed of genes that are inactive in RNA synthesis (Figure 1.26E). A typical metaphase chromosome (Figure 1.26F) contains the most highly condensed form of cellular DNA (700–1400 nm in diameter).

Techniques for viewing chromosomes have so greatly improved that it is possible to see very small deletions, duplications, inversions, and translocations. These advances, together with the routine procedure for obtaining fetal cells from a pregnant woman's amniotic fluid (*amniocentesis*), have greatly facilitated the conduct of meaningful genetic counseling. Let's look at a few of the results that have been obtained with the new methods.

Figure 1.27 is a light micrograph of a very unusual type of chromosome, the polytene chromosome seen in the salivary glands of *D. melanogaster* and other dipteran insects. The chromosomes of a meta-

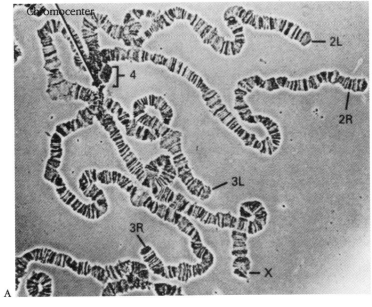

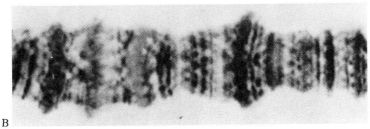

Figure 1.27 Light micrographs of polytene insect chromosomes stained to reveal a reproducible banding pattern.
(A) In the four salivary gland chromosomes (X, 2, 3, and 4) of *D. melanogaster*, a total of approximately 5000 bands can be distinguished. The tips of the metacentric 2 and 3 chromosomes are labeled (L = left arm; R = right arm). The tip of the acrocentric X chromosome is also labeled. (B) A higher power magnification of a section of chromosome C, one of the four chromosomes of the fly *R. americana*.

phase cell are stained with a dye that reacts with DNA, and the cells are then "squashed" on a microscope slide under conditions that lead to sufficient physical separation of the chromosomes of a given cell. The stain reveals a very reproducible banding pattern that is characteristic of a particular chromosome. Approximately 5000 distinct bands are present in the four chromosomes (X, 2, 3, and 4) of *D. melanogaster* (Figure 1.27A). The smallest one, number 4, appears as a dot associated with the *chromocenter*, the position at which the *centromeres* (the points by which chromosomes appear to attach to the spindle in mitosis) of all four chromosomes often appear to be fused in cells of this type. The left (L) and right (R) arms of chromosomes 2 and 3 can be seen to the right of Figure 1.27A. Figure 1.27B shows a higher power magnification of a portion of one of the four chromosomes found in another fly, *Rhyncosciara americana*. The appearance of these very large bands results from the process of selective gene amplification, whereby certain genes

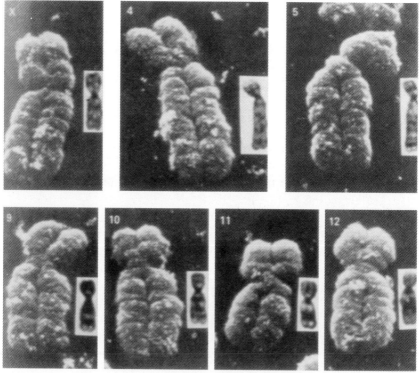

Figure 1.28 Human male chromosomes after G banding.
The chromosomes have undergone brief treatment with an enzyme that degrades protein, followed by staining with the Giemsa reagent. After this procedure, which does not stain all regions of the chromosome equally, distinctive bands appear at characteristic places. The bands enable investigators to distinguish chromosomes of similar lengths from one another. The light micrographs (*insets*) show the bands; the scanning electron micrographs show constrictions at the sites where the bands are observed.

are replicated repeatedly, generating about 1000 copies that remain closely associated in a band.

The appearance of a specific banding pattern in human chromosomes makes it possible to distinguish among those with similar sizes and shapes. Figure 1.28 shows several human chromosomes that were photographed after they had been treated briefly with a proteolytic enzyme to remove proteins and expose some regions of the DNA, then stained with Giemsa reagent (a dye that permanently stains DNA). The scanning electron micrographs show constrictions at the banding sites, while the light micrographs (insets) indicate the so-called G-bands produced by the Giemsa stain. Such techniques have permitted chromosomal abnormalities to be detected in nearly 1 percent of the human population.

The micrographic display of an entire set of chromosomes from an individual is referred to as a *karyotype*. After each pair of homologous chromosomes has been identified, pictures of the pairs are arranged side by side, in order of size. Figure 1.29 shows the karyotypes of two closely related species of deer. The *Reeves muntjac* (Figure 1.29A) has only four chromosome pairs, whereas the *Indian muntjac* (Figure 1.29B) has 23 pairs. For some time no one could understand why these two species,

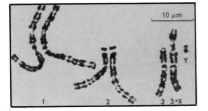

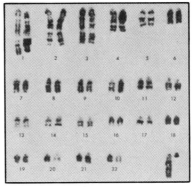

Figure 1.29   (A) The Reeves muntjac and its karyotype.
(B) The Indian muntjac and its karyotype.
After a brief digestion of proteins, the chromosomes were stained with the Giemsa reagent (G banding procedure).

which are quite similar in appearance and behavior, failed to interbreed. DNA analysis indicated that their genomes contained about the same amount of DNA. One look at the karyotypes, however, enables even a person with no cytogenetic expertise to conclude that a mating between these two species would fail to produce viable offspring. In addition to its usefulness for human genetic counseling, cytogenetics has played an important role in our efforts to preserve, at least in captivity, some of Earth's rapidly vanishing species.

Discussion Questions

1. What are the unique advantages of the phase-contrast light microscope? The scanning electron microscope? The transmission electron microscope?

2. Frye and Edidin's experiment (Figure 1.13) provided strong support for the fluid mosaic model of membrane structure. Other than lateral diffusion of the tagged proteins through the plasma membrane, what possible explanations are there for their findings?

3. Describe an experiment that would permit you to determine the rate at which the insulin receptor forms patches on the cell membrane.

4. a. How can autoradiography be used to determine the structure of a replicating DNA molecule?
    b. Describe an additional procedure in this experiment that would permit you to determine the time required for a single DNA molecule to replicate.

5. Although the intracellular SV40 DNA was shown by electron microscopy to have a chromatin-like structure, some skeptics doubted this conclusion. What additional experiment would you perform to provide more convincing evidence than that afforded by electron microscopy alone?

6. How can electron microscopy distinguish between Cairns's theta (θ) model of DNA replication and the rolling-circle model?

7. Why was such a long exposure time (five months) required in Zimm and Kavenoff's experiment that demonstrated by autoradiography that there was a single molecule of DNA in a *Drosophila* chromosome?

References

Ambros, V., L. B. Chen, and J. M. Buchanan (1975). Surface ruffles as markers for studies of cell transformation by Rous Sarcoma virus. *Proceedings of the National Academy of Sciences USA, 72:*3144.

Ash, J. S., P. K. Vogt, and S. J. Singer (1976). Reversion from transformed to normal phenotype by inhibition of protein synthesis in rat kidney cells infected with a temperature-sensitive mutant of Rous sarcoma virus. *Proceedings of the National Academy of Sciences USA, 73*:3603.

Cairns, J. (1963). The chromosome of *Escherichia coli. Cold Spring Harbor Symposium on Quantitative Biology, 28*:43.

Everhart, T. E., and T. L. Hayes (1972). The scanning electron microscope. *Scientific American, 226*:54.

Frye, L. D., and M. Edidin (1970). The rapid intermixing of cell surface antigens after formation of mouse-human heterokaryons. *Journal of Cell Science, 7*:319.

Gall, J. G. (1981). Chromosome structure and the C-value paradox. *Journal of Cell Biology, 91*:3S.

German, J. (1970). Studying human chromosomes today. *American Scientist, 58*:182.

Goldstein, J. L., R. G. W. Anderson, and M. S. Brown (1979). Coated pits, coated vesicles, and receptor-mediated endocytosis. *Nature, 279*:679.

Green, M. H., H. I. Miller, and S. Hendler (1971). Isolation of a polyoma-nucleoprotein complex from infected mouse cell cultures. *Proceedings of the National Academy of Sciences USA, 68*:1032.

Griffin, F. M., Jr., J. A. Griffin, and S. C. Silverstein (1976). Studies on the mechanism of phagocytosis. II. The interaction of macrophages with anti-immunoglobulin IgG-coated bone marrow-derived lymphocytes. *Journal of Experimental Medicine, 144*:788.

Griffith, J. D. (1975). Chromatin structure: deduced from a minichromosome. *Science, 187*:1202.

Harrison, C. J., M. Britch, T. D. Allen, and R. Harris (1981). Scanning electron microscopy of the G-banded human karyotype. *Experimental Cell Research, 134*:141.

Heuser, J. (1981). Quick-freeze, deep-etch preparation of samples for 3-D electron microscopy. *Trends in Biochemical Science, 6*:64.

Hourcade, D., D. Dressler, and J. Wolfson (1973). The nucleolus and the rolling circle. *Cold Spring Harbor Symposium on Quantitative Biology, 38*:537.

Kahn, C. R., K. L. Baird, D. B. Jarrett, and J. S. Flier (1978). Direct demonstration that receptor crosslinking or aggregation is important in insulin action. *Proceedings of the National Academy of Sciences USA, 75*:4209.

Kavenoff, R., L. C. Klotz, and B. H. Zimm (1973). On the nature of chromosome-sized DNA molecules. *Cold Spring Harbor Symposium on Quantitative Biology, 38*:1.

Kleinschmidt, A. K., D. Lang, D. Jacherts, and R. K. Zahn (1962). Preparation and length measurements of the total deoxyribonucleic acid content of T_2 bacteriophages. *Biochim. Biophys. Acta, 61*:857.

Koenig, S., H. E. Gendelman, J. M. Orenstein, M. C. Dal Canto, G. H. Pezeshkpour, M. Yungbluth, F. Janotta, A. Aksamit, M. A. Martin, and A. S. Fauci (1986). Detection of AIDS virus in macrophages in brain tissue from AIDS patients with encephalopathy. *Science, 233*:1089.

Morgan, C., and H. M. Rose (1968). Structure and development of viruses as observed in the electron microscope. *Journal of Virology, 2*:925.

Pease, D. C., and K. R. Porter (1981). Electron microscopy and ultramicrotomy. *Journal of Cell Biology, 91*:2875.

Rogers, A. W. (1979). *Techniques of Autoradiography* (3rd ed.). New York: Elsevier/North-Holland.

Salzman, N. P., G. C. Fareed, E. D. Sebring, and M. M. Thoren (1973). The mechanism of SV40 DNA replication. *Cold Spring Harbor Symposium on Quantitative Biology, 38*:257.

Scheer, U., and H. Zentgraf (1978). Nucleosomal and supranucleosomal organization of transcriptionally inactive rDNA circles in *Dytiscus* oocytes. *Chromosoma, 69*:243.

Schlessinger, J., Y. Shechter, M. C. Willingham, and I. Pastan (1978). Direct visualization of binding, aggregation, and internalization of insulin and epidermal growth factor on living fibroblastic cells. *Proceedings of the National Academy of Sciences USA, 75*:2659.

Singer, S. J., and G. L. Nicolson (1972). The fluid mosaic model of the structure of cell membranes. *Science, 175*:720.

Sowers, A. E., and C. R. Hackenbrock (1981). Rate of lateral diffusion of intramembrane particles: measurement by electrophoretic displacement and rerandomization. *Proceedings of the National Academy of Sciences USA, 78*:6246.

Spencer, M. (1982). *Fundamentals of Light Microscopy*. New York: Cambridge University Press.

2

Purification and Properties of Protein Kinase

2.1 Protein Purification Techniques

2.1.1 The Importance of Protein Purification

The purification and analysis of specific proteins probably consume more laboratory hours per person than any other endeavor in modern biology. Why is this the case? The reason may be that there are several thousand different proteins, each with its unique and fascinating properties. Just as cells are the building blocks of all living things, with the single exception of viruses, proteins are the life of every cell. They determine the cell's structure as well as most of its functions — metabolism, growth, division, control of gene function. Thus biologists in fields as diverse as cell biology, molecular biology, genetics, developmental biology, evolution, and medicine are all concerned with the biochemical properties of proteins. The importance of proteins in all of these fields will become clear shortly.

Let us begin with some of the most important procedures that are commonly used to purify proteins, emphasizing the principles rather than the detailed procedures. We shall then focus on one typical class of proteins, the protein kinases. These enzymes catalyze the transfer of a phosphate from ATP to a protein. First we shall see how the efforts of biochemists to purify what was thought to be a single enzyme led to the realization that there are actually many kinds of protein kinases, each with an important role in the regulation of the cell's functions. Not only could the protein kinase activities in the cell be regulated by small molecules such as cyclic AMP (adenosine monophosphate), but the activities of many other enzymes proved to be controlled by the activities of specific protein kinases. As we explore the methods for determining the identity of a protein that is encoded by a particular gene, we shall learn how it was discovered that certain protein kinases play a critical role in the conversion of a normal cell into a cancer-causing (tumorigenic) cell. We shall then see how studies on the amino acid sequences of proteins led to current theories concerning the way the primordial prokaryotes evolved into today's diverse eukaryotic world.

2.1.2   Cell Fractionation

The first step in the purification of a protein generally entails a procedure known as *cell fractionation*. One selects a source that is rich in the desired protein, perhaps a particular organ or an even more homogeneous collection of cells, such as a clonal population grown in cell culture. A typical flow diagram for a cell fractionation scheme is seen in Figure 2.1.

The cells are broken open by one of a number of methods. These include sonication with high-frequency sound waves; *osmotic* or *hypotonic shock*; and homogenization, which can be effected by a glass rod that fits snugly inside a glass tube (a Dounce homogenizer). After unbroken cells and large clumps of connective tissue or other unwanted debris have been removed by centrifugation or filtration, the suspension is subjected to a series of centrifugations at progressively increased centrifugal forces or for progressively longer times, or both. This procedure, referred to as *differential centrifugation*, causes organelles and other particulate structures to sediment to the bottom of the centrifuge tube, forming a *pellet*. The fluid fraction is referred to as the *supernatant*, or more often as the *sup* or *supernate*.

The *force-times* (centrifugal force multiplied by time of centrifugation) that cause various cellular components to form a pellet are indicated in Figure 2.1. Also shown is a diagram of a rotor in an ultracentrifuge, a machine capable of generating forces well in excess of the $100,000 \times g$ needed to pellet small particles such as ribosomes. After the pellet is formed, the remaining fluid, called the *cytosol*, contains the soluble proteins and other cytoplasmic molecules. Thus, with a variety of centrifuges that generate different centrifugal forces, one can obtain a cell fraction that is enriched in a particular protein.

2.1.3   Column Chromatography

The next step in protein purification often involves one or more types of column chromatography. Three of the most common types are ion-exchange, gel-filtration, and affinity chromatography.

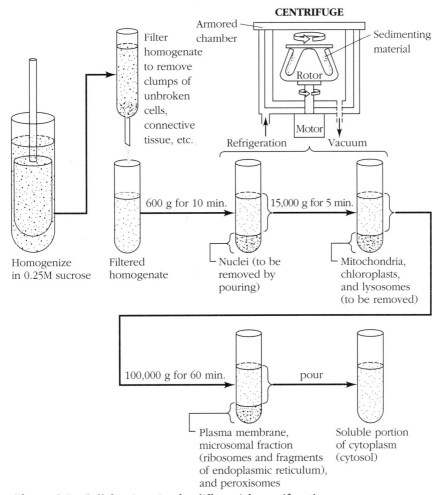

Figure 2.1 Cell fractionation by differential centrifugation.

Ion-exchange resins are solids that can exchange ions with aqueous solutions. The resins are covalently linked to either positively or negatively charged chemical groups, creating anion-exchange or cation-exchange columns, respectively. For protein separations, the most common resins consist of cellulose beads, such as phosphocellulose (negatively charged) and diethylaminoethyl (DEAE) cellulose (positively charged). Proteins can be separated from one another on the basis of their net ionic charge, which affects their electrostatic attraction to the charged resin.

The protein solution is passed through a column containing the ion-exchange resin, and proteins that have a charge opposite that of the resin bind near the top (Figure 2.2). A solution (*eluant*) containing

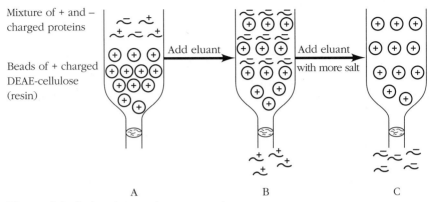

Figure 2.2 Ion-exchange chromatography.
(A) A mixture of positively and negatively charged proteins is added to an ion-exchange resin (beads), positively charged DEAE-cellulose, in a column. (B) The positively charged proteins are not bound very tightly to the positively charged resin, and so are eluted more quickly by the eluant. (C) As the salt concentration of the eluant increases, the negatively charged proteins are eventually eluted from the positively charged resin.

increasing concentrations of salt is then used to elute, or wash off, the bound proteins from the resin. As the salt becomes sufficient to displace a protein from the solid matrix, the protein travels down the column and eventually into a waiting test tube. The procedure can be automated with sophisticated electronic equipment—fraction collectors, pumps that control the flow rate, monitors that measure and control the pH and ionic strength of the eluant, recorders that enable one to detect which tubes contain the eluted proteins.

Gel filtration is another effective method for purifying a specific protein. In this case, size rather than charge provides the basis of the separation. The column is packed with uncharged porous beads, such as Sephadex (Figure 2.3). Protein molecules that are small enough can enter the beads through the pores, and are thus greatly retarded in their elution from the column. The proteins too large to fit into the pores travel in between the beads and are eluted more rapidly. This type of column is also referred to as a *molecular sieve*, and it is frequently used to change the ionic composition of a solution containing macromolecules, such as proteins and nucleic acids. The macromolecules are simply eluted with a solvent containing the desired ions, while the original ions are captured by the porous beads. In this respect gel filtration accomplishes the same thing as dialysis, but much more rapidly.

Affinity chromatography is by far the most efficient type of protein fractionation procedure, in that any one of the proteins in a complex mixture can be brought to a high degree of purity after a single run through the column. This process takes advantage of the high affinity of an enzyme for its *substrate*, its *competitive inhibitor*, or some other

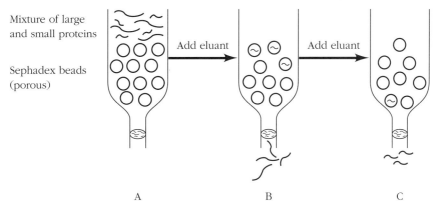

Figure 2.3 Gel-filtration chromatography (molecular sieving).
(A) A mixture of large and small proteins is added to a gel filtration column
with its porous beads. (B) The small proteins enter the porous beads, but the
large proteins cannot and thus are eluted sooner. (C) As sufficient eluant
flows through the column, the small proteins are eventually eluted.

molecule. The basis of this procedure is essentially as follows: a substrate designated for a specific protein is covalently bonded to tiny beads
(such as *agarose*, a naturally occurring polysaccharide), which are
placed in a column (Figure 2.4). The mixture of proteins is then washed
through the column. Only proteins capable of binding to the substrate
will adhere to the column. After the unbound proteins are eluted, the

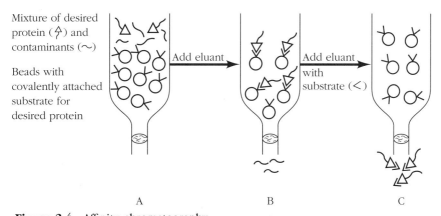

Figure 2.4 Affinity chromatography.
(A) A mixture of proteins is added to an affinity chromatography column
containing beads to which a substrate for the desired protein (enzyme) is
covalently attached. (B) The desired protein (◁~) binds tightly to the beads
while the contaminating proteins (~) are eluted by the eluant. (C) The
desired protein is eluted from the beads upon addition of a high concentration
of substrate (<) molecules in the eluant. The added substrate molecules
displace the protein from the beads by competing for binding sites on the
protein.

bound ones can be eluted with a solution containing a high substrate concentration. The soluble substrate disrupts (by competition) the binding between the protein and the resin-bound substrate.

2.2   Purification of Nuclear Protein Kinase

The covalent linkage of phosphate to amino acid residues in proteins, a *phosphorylation* reaction catalyzed by enzymes called *protein kinases,* was first demonstrated by F. Lipmann and P. A. Levene in 1932. The significance of this reversible modification of proteins in eukaryotic cells became increasingly apparent with the pioneering studies of E. W. Sutherland, E. G. Krebs, and others since the 1950s. Protein phosphorylation plays a key role in the regulation of metabolic processes and of cellular growth in virtually all organisms.

We now know that protein kinases can either activate or inactivate enzymes by catalyzing their phosphorylation, and there are dozens of protein kinases in virtually all types of cells. One classification scheme for these enzymes is based on the target protein's amino acid acceptor of the phosphoryl group, which can be an alcohol group, as in serine and threonine; a phenolic group, as in tyrosine; or a nitrogen-containing group, as in histidine and lysine. Each of these classes of protein kinase can be further subdivided on the basis of regulatory agents that affect their activity via the *allosteric site*, a region of the enzyme that is separate from the active site. There are many such agents, some of them molecules that have a wide variety of structures, such as cyclic nucleotides (cyclic AMP, for example), Ca^{++}, diacylglycerol, and peptide hormones. Thus, while all the active sites of the protein kinases recognize ATP, the source of the phosphoryl group, the allosteric regulatory sites are quite diverse.

2.2.1   Key Terms

Obviously the discovery that a wide variety of protein kinases exist in any one cell was possible only through the development and painstaking application of many biochemical techniques that resulted in their purification. Before we see how this was actually done, a few key terms in the field of enzymology must be defined.

During the purification procedure, it is important to keep an account of the *enzyme activity* that is present in each of the usually large number of fractions. Just as the unit for weight is the gram, the unit for enzyme activity is called the *enzyme unit* (U). One U may be defined as the amount of activity that converts one micromole (μmol) of substrate into product per second. This activity is not to be confused with *specific activity*, which can be defined as the ratio of enzyme activity to the total mass of protein in the sample. Whereas enzyme units measure the quantity of enzyme present, the specific activity is an indicator of enzyme purity irrespective of its quantity. The significance of these essential terms will become clear as we proceed.

2.2.2.   The Enzyme Assay Method

For our case study on protein kinase purification, we shall examine the work of Dr. Françoise Farron-Furstenthal, a Viennese biochemist who dedicated most of her scientific career to the study of this class of enzymes. In 1975, while at the Stanford Research Institute in California, Dr. Farron-Furstenthal investigated the subcellular distribution of protein kinases in adult and fetal rat liver and in a rat liver tumor (hepatoma). Her results revealed three- to four-fold greater percentage of the total activity in the nuclei of growing liver cells than in the nuclei of adult liver. This discovery and related findings led the way to the identification of tumor-specific protein kinases.

A good enzyme assay must be relatively easy to do, as it is generally performed on hundreds of samples each time the enzyme is purified from a crude homogenate. The assay must also be farily inexpensive and precise. Figure 2.5 shows a flow diagram of Dr. Farron-Furstenthal's assay for protein kinases.

Note that the reaction uses radiolabeled ATP, with the γ (or third) phosphate containing ^{32}P. This is the phosphorus that is transferred to the protein substrate by protein kinase. After each reaction is stopped by the addition of unlabeled ("cold") ATP and ice-cold trichloroacetic acid (TCA), a protein carrier, bovine serum albumin (BSA), is added and the mixture is placed on ice for 30 minutes. During this time a precipitate of macromolecules forms, including the BSA and the newly labeled protein. The role of the protein carrier is to ensure complete precipitation of the radiolabeled protein. Each reaction mixture is then filtered through a Whatman glass fiber disc, which separates the ^{32}P-protein product from the ^{32}P-ATP. The filters, to which the ^{32}P-protein is bound, are thoroughly washed, dried, and placed in vials. After a solvent such as toluene (a *scintillation fluid*) is added, the samples are ready to be analyzed for ^{32}P-protein. Under appropriate conditions of assay, the amount of this radiolabeled protein is directly proportional to the amount of enzyme that was present in the assay.

The vials are placed in a liquid scintillation counter, such as the one shown in Figure 2.6, programmed to analyze the particular radioisotopes that are present. Each isotope, commonly ^{32}P, ^{35}S, ^{14}C, and ^{3}H, emits β particles (electrons) at a characteristic energy. These particles

1. Incubation:

$$\gamma^{32}\text{P-ATP} + \text{Protein} \xrightarrow[\text{10 minutes, 30°C}]{\text{Enzyme}} \text{}^{32}\text{P-protein} + \text{ADP}$$

2. Stop reaction:

Add "cold" ATP + TCA (acid) + BSA (carrier protein).
Store 30 minutes on ice. ^{32}P-protein precipitates.

3. Filter:

Pour reaction mixture through Whatman glass fibers.
Wash filters with cold TCA, then 95% ethanol.
^{32}P-protein sticks on filters.

4. Count radioactivity on filters:

Place dry filters in vials containing scintillation fluid
(toluene + fluors). Place vials in liquid
scintillation counter.

Figure 2.5 Assay for protein kinases.

strike organic *fluors* that are dissolved in a scintillation fluid, causing the emission of light with an intensity that depends on the energy of the β particles. The amount of radioisotope in the sample is proportional to the number of flashes of light, which are detected by the scintillation counter at the programmed intensity and printed out in terms of counts per minute (CPM). The experimenter can return to the machine after lunch or a good night's sleep, depending on the number of samples, and collect the data, which, if necessary, can be graphed by a computer.

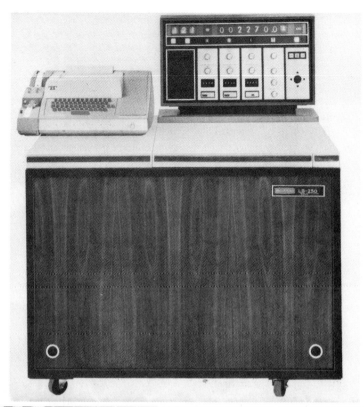

Figure 2.6 Liquid scintillation counter.
Radioisotopes (^{32}P, ^{35}S, ^{14}C, ^{3}H, etc.) emit β particles (electrons), which strike fluors and cause emission of light. Amount of radioactivity (counts per minute, CPM) is proportional to number of flashes of light.

2.2.3 Experimental Findings

Returning now to the experiment, we find that protein kinase activities did not differ appreciably in the unfractionated homogenates of adult and fetal liver and the hepatoma, ranging in value from 18.7 to 31.0 per gram of fresh tissue (Table 2.1). The subcellular distribution, however, did vary significantly. Activity was three to four times greater in the nuclei purified from the rapidly growing fetal and tumor tissues than in the nuclei from adult liver. These results fitted nicely with earlier observations of an increased level of phosphorylation of nuclear proteins in actively growing tissues and cells in comparison with nondividing cells.

Was the increased activity of nuclear protein kinase due to the presence of a greater number of enzyme molecules or to the existence of a new protein kinase in the tumor cells? This question was clearly answered by a single experiment in which ion-exchange chromatography was used to resolve nuclear protein kinase activities. Nuclear protein from either adult liver or hepatoma cells was loaded onto two identical columns containing DEAE cellulose. The columns were eluted simultaneously with a solution of linearly increasing salt concentration dispensed from a single gradient maker. Whereas the normal adult liver nuclei exhibited three peaks of protein kinase activity, the hepatoma extract resolved into five peaks. Figure 2.7 shows clearly that the liver tumor nuclei possessed at least two protein kinases (peaks II and V) that

Table 2.1. Subcellular distribution of protein kinase activity in the Morris hepatoma 7288C, adult and fetal liver of the rat.

Cell fraction	Enzyme activity (mμmol/g/min)	Distribution of enzyme activity (percent)	Specific activity (mμmol/mg protein)
Adult Liver			
Homogenate	21.6	100	0.09
Postnuclear supernate	16.8	78	0.10
Purified nuclei	0.8	3.7	0.086
Hepatoma			
Homogenate	31.0	100	0.142
Postnuclear supernate	22.0	71	0.15
Purified nuclei	4.25	13.7	0.21
Fetal Liver			
Homogenate	18.7	100	0.095
Postnuclear supernate	12.3	66	0.092
Purified nuclei	2.0	10.7	0.185

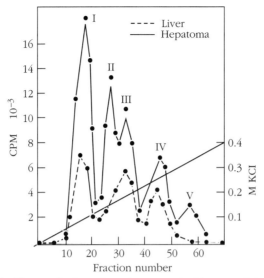

Figure 2.7 The resolution of nuclear protein kinase activity by DEAE-cellulose chromatography.

were not present in the nuclei of normal liver cells. These enzymes could conceivably play a key role in the abnormal growth properties of cancer cells.

Further purification of the nuclear protein kinases was achieved by affinity chromatography. A flow diagram of the purification scheme is shown in Figure 2.8. Purified nuclei from 15g rat liver were suspended in 0.5M NaCl, pH 7.5, and digested for 30 minutes at 20°C with DNAase and RNAase to hydrolyze the nucleic acids. The suspension was centrifuged at 7000 × g for 10 minutes. The supernate was then dialyzed for 16 hours against 100 volumes of a buffer containing 0.1M NaCl, pH 7.5. The precipitate that formed upon this reduction in salt concentration was removed by centrifugation. The supernatant contained the protein kinase activity.

It should be noted that this simple procedure of changing the salt concentration is a very useful step in protein purification. As Figure 2.9 indicates, neutral salts have pronounced effects on the solubility of different proteins. Many proteins have increased solubility at low salt concentrations, a phenomenon called *salting in.* At high and extremely low salt concentrations, proteins may be completely precipitated, or *salted out.* Since proteins vary in their response to neutral salt concentrations, salting in and salting out are important procedures in the separation of protein mixtures.

Returning to our flow scheme (Figure 2.8), we see that the next step in the purification of nuclear protein kinases involved a casein-Se-

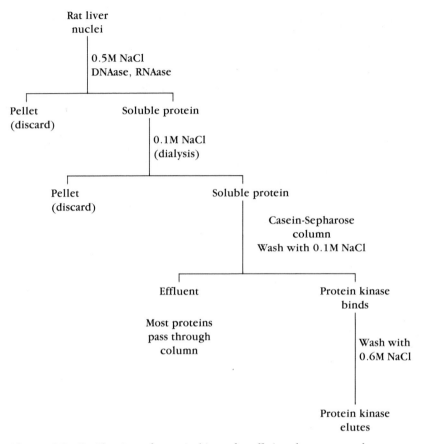

Figure 2.8 Purification of protein kinase by affinity chromatography.

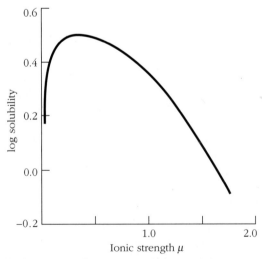

Figure 2.9 Salt effects on carbon monoxide-hemoglobin.

pharose column. Casein is an inexpensive, commercially available milk protein that serves as a substrate for phosphorylation by protein kinases. When it is linked covalently to Sepharose beads, the resulting resin has a high affinity for protein kinases. After the soluble protein was added to the casein-Sepharose column, the column was eluted first with 0.1M NaCl and then with 0.6M NaCl. Fractions were collected and assayed for protein kinase activity.

Several interesting points can be seen from the data obtained (Table 2.2). First, it is apparent that the total enzyme activity observed with casein as a substrate was greater than 100 percent after extraction with 0.5M NaCl. How is this possible? It should be realized that crude cell extracts contain all sorts of inhibitors. Some act on the enzyme itself and others interfere with the substrates or ions in the reaction mixture, making the conditions for the reaction suboptimal. Second, we note that specific activity has increased from 0.58 with the purified nuclei to 43.5 with the 0.6M NaCl effluent from the affinity column. This represents a 75-fold purification of the nuclear protein kinases, with a yield of 79 percent of the original activity, an excellent result in view of the simplicity of the procedure.

Finally, the assays indicate that nuclear protein kinase activity has been cleanly separated from *endogenous* protein substrates; that is, protein substrates present in the enzyme sample. This is evident in the data in the columns showing the activities with endogenous substrate and additional protein substrate; that is, *casein*. Protein kinase activity was negligible with endogenous substrate in the effluent at 0.6M NaCl. This effective separation had not been achieved by other chromatographic procedures. It then became possible for researchers to investigate whether the separate peaks of protein kinase (evident in Figure 2.6)

Table 2.2.   Purification of protein kinase from rat liver nuclei.

	Volume	Substrates (total units)		Yield (percent)	Total protein (mg)	Specific activity[a]
		Endogenous	Casein			
Purified nuclei	10.0	6.0	13.2	100	22.7	0.58
0.5M NaCl extract	8.0	7.1	17.0	129	18.0	0.94
Proteins soluble in 0.1M NaCl	5.3	4.4	12.1	92	9.7	1.25
Casein-Sepharose effluent at 0.1M NaCl (substrate)	18.0	0.23	0.7	—	7.3	—
Casein-Sepharose effluent at 0.6M NaCl (enzyme)	3.3	nil	10.4	79	0.24	43.5

[a]Based on activity measured with casein as substrate.

exhibited preferences for specific protein substrates. The identification of the natural protein substrate for each protein kinase remains an important goal in this field of research.

2.2.4   When Is a Protein Pure?

With more than a thousand different proteins present even in prokaryotic organisms, how can one know whether a desired protein has been purified to homogeneity, and why is it important to find out? The ultimate goal of investigation into a protein is to understand the relationship of its structure and function. Even trace impurities of other proteins can cause considerable difficulties, for example, in efforts to determine the amino acid sequence or to obtain crystals for three-dimensional structure analysis by X-ray crystallography. Of course, any protein impurity makes it difficult to know which protein has the activity of interest, and many types of competitive activities in impure preparations of enzymes can completely confuse interpretation of the data.

The current method of choice for determining both the purity and the molecular weight of a protein is gel electrophoresis. This technique is based on the fact that dissolved molecules that are subjected to the force of an electric field move at a speed determined by the ratio of their charge to their mass. However, most proteins have very similar charge-to-mass ratios, even if they differ in shape and mass. This problem is overcome by the use of gels prepared from agarose, a naturally occurring polysaccharide, or from *polyacrylamide*, a more stable synthetic polymer. The size of the pores in these gels can be varied during their preparation, and these pores limit the rate at which macromolecules can move through the gel. It should be noted that the pores in these gels, unlike those used in filtration chromatography (Figure 2.3), are between the gel particles and not inside them. Thus the smaller the protein, the more rapidly (easily) it passes through the gel matrix toward the pole that has an opposite charge (Figure 2.10). The gel also serves to retard diffusion of the separated proteins after the current is turned off.

In determining the purity of a protein preparation, it is often of advantage to *denature* the protein into its polypeptide subunits. For this purpose, a strong polyanionic detergent such as *sodium dodecyl sulfate* (SDS) is most often the agent used. Approximately one molecule of SDS binds to each amino acid, creating strong electrostatic repulsion within the protein and denaturing it completely into negatively charged polypeptide subunits. Even proteins that are inadvertently stuck together will separate when they are exposed to SDS. Thus a protein aggregate that appears to be a single pure protein when it is analyzed by polyacrylamide gel electrophoresis (PAGE) will be seen as a mixture of polypeptides when it is run under denaturation conditions (SDS-PAGE).

Because many proteins consist of more than one polypeptide chain, the experimenter is faced with the problem of determining

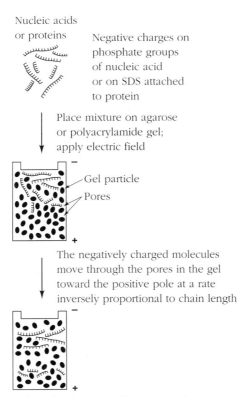

Nucleic acids
or proteins

Negative charges on
phosphate groups
of nucleic acid
or on SDS attached
to protein

Place mixture on agarose
or polyacrylamide gel;
apply electric field

Gel particle

Pores

The negatively charged molecules
move through the pores in the gel
toward the positive pole at a rate
inversely proportional to chain length

Figure 2.10   Principles of gel electrophoresis in the separation of proteins
and nucleic acids.

whether the appearance of more than one protein band after SDS-PAGE
is caused by impurities in the preparation. Using appropriate protein
markers that have well-characterized molecular weights, one can readily
determine the mass of each of the proteins in the gel. The total mass of
the polypeptides seen by SDS-PAGE should add up to that of the protein
estimated under nondenaturation conditions. If it does not, contaminants are certainly present.

Three general types of apparatus can be used for gel electrophoresis. The gel can be formed in a glass tube or on a "slab," so that the gel is
sandwiched between two thin glass sheets. Even more simply, the gel
can be poured onto a horizontal surface, or "flat bed." An example of
tube gel electrophoresis is shown in Figure 2.11. The tubes are inserted
between two buffer chambers, one containing the positive and the other
the negative electrode. Thus the only electrical contact between the two
chambers is through the gel. When current is applied to the buffer
compartments, the proteins begin to migrate at rates inversely proportional to their masses. When the desired length of time has elapsed, the
current is turned off and the tubes are removed. The gels are then
removed from their glass containers and stained, usually with Coomassie
blue or with a silver stain. When the gel has been properly stained, a

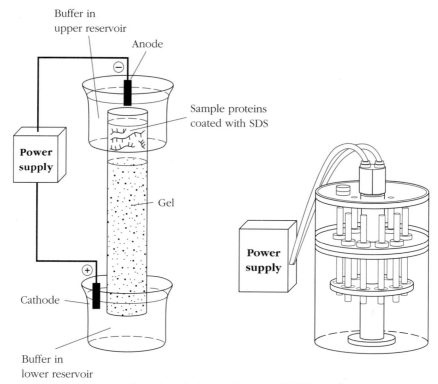

Figure 2.11 Polyacrylamide gel electrophoresis (PAGE) apparatus. Proteins are denatured by treatment with sodium dodecyl sulfate (SDS), which makes them negatively charged. The protein sample is layered on top of the gel at the negative pole. The current is turned on and the proteins migrate through the gel toward the positive pole at rates inversely proportional to their molecular weights.

series of bands appears, distinguishing and defining the different species of proteins. The intensity of the stain is directly related to the amount of protein present.

Many other methods can be used to assess the purity of a protein. High-performance liquid chromatography (HPLC) is a very versatile and powerful technique for the purification and assay of proteins, nucleic acids, and other biomolecules. HPLC uses the same types of chromatography described earlier; that is, gel exclusion, ion exchange, and affinity resins. However, greater and faster resolution is achieved through the use of column materials that consist of more finely divided particles that are strong enough to withstand pressures as high as 10,000 pounds per square inch without undergoing changes in structure. The high pressure permits very rapid flow rates, and thus rapid elution of the sample. However, the high cost of the apparatus for HPLC ensures that gel electrophoresis will remain the most frequently used procedure.

Another important method, both for protein purification and for quantitative evaluation of purity, is electrofocusing. This technique employs gel electrophoresis, but is dependent solely on the protein's net charge, not on its size. The stabilizing gels, again usually polyacrylamide or agarose, are prepared to include a mixture of small charged organic molecules, called *ampholytes*. When current is applied to the gel, the ampholytes quickly form a linear pH gradient. The sample proteins move through the pH gradient until they reach a position at which the pH of the gradient is equal to the isoelectric point (pI) of the protein. This is the point in the pH gradient at which the protein stops moving because it has lost all of its net charge. Two proteins with nearly identical molecular weights, differing only by a single phosphate group, can be resolved by this isoelectric focusing technique. With resolutions as refined as 0.01 pH units, electrofocusing is a highly sensitive assay for protein impurities and those modified forms of the same protein that involve a change in net charge.

2.3   Control of Enzyme Activity

Two enzymological experiments by Dr. Edwin G. Krebs and his collaborators helped to establish the central role of a protein kinase in the regulation of blood sugar levels. This work helps us to appreciate the real value of a purified enzyme.

A protein kinase that exhibits a complete dependence on cyclic AMP for its activity, obtained from rabbit skeletal muscle, was partially purified (300-fold). Earlier work had indicated that cyclic AMP is an intracellular mediator of numerous hormonal responses, but the mechanisms by which this nucleotide carries out its role remained unclear. With the purified cyclic AMP-dependent protein kinase, Krebs and his colleagues demonstrated that this enzyme can activate another kinase — namely, phosphorylase kinase — thereby regulating its activity in the breakdown of glycogen.

2.3.1   Protein Kinase Activation by Cyclic AMP

Krebs and his associates first demonstrated that the purified protein kinase catalyzed the transfer of ^{32}P from $\gamma^{32}P$-ATP to a protein substrate, casein, in a reaction that exhibited a complete dependence on the presence of cyclic AMP. With excess casein, the reaction rate was linear over a wide range of enzyme concentrations, and thus served as a convenient assay system for the protein kinase.

As Figure 2.12 shows, the phosphorylation of casein was very sensitive to low concentrations of cyclic AMP. The apparent K_m (Michaelis constant) value for cyclic AMP was 1×10^{-7} M. This means that when the concentration of cyclic AMP is 10^{-7} M, the rate of the reaction proceeds at one-half the maximal rate for a limiting concentration of enzyme (see below). Concentrations that are close to the K_m value are quite likely to be of physiological significance, with relatively small

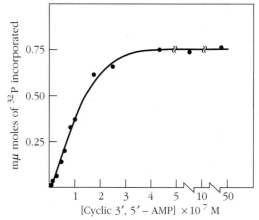

Figure 2.12 Effect of cyclic AMP on casein phosphorylation by protein kinase. Reaction mixtures contained a fixed concentration of protein kinase (0.113 mg per ml) and were incubated for 10 minutes at 30°C. Reactions were initiated by the addition of γ^{32}P-ATP and terminated by the addition of 10% tricholoroacetic acid and carrier protein (bovine serum albumin).

changes producing significantly different reaction rates (as Figure 2.12 indicates). This is obviously not the case when substrate concentrations are so large that changes do not affect the reaction rate.

The Michaelis constant, K_m, is related to substrate concentration, S, by the equation

$$r = \frac{V_{max} \cdot S}{K_m + S}$$

where r = the rate of reaction and V_{max} = the maximal rate at excess substrate.

When S and K_m both equal 10^{-7} M,

$$r = \frac{V_{max} \cdot 10^{-7}}{10^{-7} + 10^{-7}} = \tfrac{1}{2} V_{max}$$

2.3.2 Activation of Phosphorylase Kinase by Protein Kinase

Phosphorylase kinase is essentially inactive when it is purified from skeletal muscle. However, when this enzyme was incubated with the purified cyclic AMP-dependent protein kinase plus ATP, it became activated (Figure 2.13). Other experiments showed that this activation

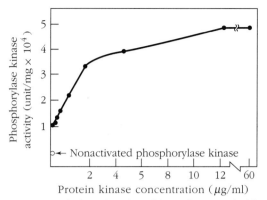

Figure 2.13 Activation of phosphorylase kinase by protein kinase.
Nonactivated phosphorylase kinase (0.3 mg per ml) was incubated for 1
minute at 30°C in reaction mixtures containing ATP, cyclic AMP, and varying
amounts of purified protein kinase.

was accompanied by phosphorylation of the enzyme phosphorylase kinase. Krebs and his associates thus concluded that cyclic AMP can activate protein kinase directly, via interaction with an *allosteric site*, and
thereby indirectly activate phosphorylase kinase, and presumably many
other enzymes, via ATP-dependent phosphorylation.

2.3.3   The Relation of Protein Kinase Activity to Blood Sugar Level

To see just one of the critical roles played by protein kinases in
animal physiology, consider the regulation of blood sugar levels. The
complexity of this control system is apparent from one look at Figure
2.14. Before pressing the panic button, let's take a brief tour through this
biochemical cascade. The relevance of enzymology to everyday life will
soon become more obvious.

Our bodies often demand a sudden burst of energy (ATP). For
instance, you may have to duck away from a pot that has been hurled at
your head by a "friend" who has been excessively annoyed in the
kitchen. This fight-or-flight response is controlled by the adrenal gland,
which secretes a hormone called epinephrine, also known as adrena-
line, into the bloodstream. Epinephrine binds to a specific receptor site
on the outer surface of the cell membrane. This binding results in a
conformational change in the epinephrine receptor, which activates
adenylate cyclase, an enzyme bound to the inner surface of the cell

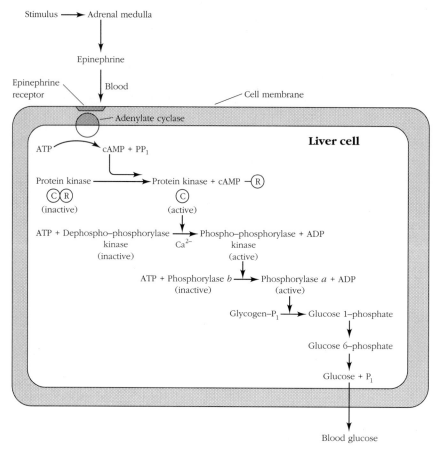

Figure 2.14   Protein kinase and the regulation of blood sugar.

membrane. In its active form, adenylate cyclase is able to convert ATP into cyclic AMP (cAMP). The cAMP binds to the regulatory subunit of protein kinase, converting this enzyme to its active form. Thus activated, protein kinase catalyzes the phosphorylation and activation of phosphorylase kinase, which in turn activates phosphorylase. Phosphorylase catalyzes the breakdown of glycogen to yield glucose-1-phosphate, which enters the *glycolytic pathway* and results in the production of large amounts of ATP.

As long as epinephrine is secreted into the blood, the adenylate cyclase system remains activated, thus maintaining a high level of cAMP and blood sugar. When epinephrine secretion ceases, however, cAMP is no longer formed. Protein kinase then reverts to its inactive form, thereby inactivating phosphorylase kinase, so that glycogen is synthesized rather than broken down.

The ready reversibility of glycogen synthesis and degradation is a very important physiological process. I have discussed only a small portion of what is known about it. If I were to say more, it would be necessary to discuss at least two other vital hormones, insulin and glucagon. The horrors of diabetes provide dramatic evidence of the need for accurate control of the blood sugar level. The rewarding findings of enzymologists working with protein kinases and many other enzymes paved the way for our eventual understanding of these complex physiological interactions.

2.4 Identification of the Protein Encoded by a Specific Gene

Sometimes a particular gene is known to play a critical role in the etiology of a disease, but the nature of the gene product is a complete mystery. Like Sherlock Holmes, modern biologists use a great number of rather simple techniques to solve such enigmas. Of course, once they are solved, they don't seem at all difficult. The work of Marc Collett and Raymond Erickson (1978) led to the first identification of the enzymatic activity of a viral gene product that was known to be responsible for the conversion (transformation) of a normal cell into a cancerous (tumorigenic) cell.

2.4.1 The *src* Gene in Avian Tumor Viruses and Cancer Cells

The Rous sarcoma virus (RSV), discovered by Peyton Rous in 1911, is one of a large family of *retroviruses*, cancer-causing viruses that have RNA for a genome and *reverse transcriptase* encapsulated within their membranous shells. Reverse transcriptase has the unique feature of copying viral RNA backward into DNA after the virus infects an animal cell. This enzyme is one of the four gene products encoded by RSV. Two others are found as structural proteins in the viral particles. The remaining gene, designated *src* (pronounced *sark*), was known by 1970 to be responsible for the ability of RSV to cause sarcomas in chickens and to transform infected animal cells in culture. *Sarcomas* are tumors that originate from connective tissue and muscle (see Chapter 4). The race began in many laboratories to characterize the protein encoded by the *src* gene.

In 1977 the Erickson laboratory reported that the product of the *src* gene had a molecular weight (M_r) of 60,000, and it was thus designated p60*src* (that is, protein of 60,000 d encoded by the *src* gene). Within a year, Collett and Erikson demonstrated that protein kinase activity was associated with the RSV *src* gene product, placing this enzyme on center stage with respect to the drama that quickly unfolded concerning the molecular basis of cancer.

2.4.2 The Basis of the Immunoprecipitation Assay

The major technique employed in this project entailed the *immunoprecipitation* of tumor-cell-specific proteins, as outlined in Figure 2.15. Tumors were induced in New Zealand rabbits by injection of

1.
Rabbits with ASV tumors $\longrightarrow$ Antiserum
(antibodies to tumors)

2.
Cell extract $+$ Antibodies $\xrightarrow[4°C]{30 \text{ min}}$ Ag-Ab $\xrightarrow{\text{Staph}}$ Ag-Ab-A
(Tumor antigens $=$ Ag) (Ab) (protein A) (precipitate)

3.
Suspend Ag-Ab-A in reaction mixture $+\gamma^{32}$P-ATP; incubate 10 min at 30°C. Protein kinase in Ag-Ab-A adds ^{32}PO$_4$ from γ^{32}P-ATP to proteins in Ag-Ab-A.

4.
Dissolve Ag-Ab-A in detergent (SDS) $+$ 10% mercaptoethanol; heat at 95°C for 1 min; centrifuge; assay supernatant proteins by PAGE and autoradiography.

Figure 2.15 Procedure for the immunoprecipitation assay.

purified virus. The actual virus used was the Schmidt-Ruppin (SR) strain of avian sarcoma virus (ASV), designated SR-ASV, a close relative of RSV. *Antiserum* from these tumor-bearing rabbits was designated TBR serum. The TBR serum contained antibodies that bound to and precipitated SR-ASV structural proteins as well as the viral *src* gene product of M_r 60,000. The reason they did so lies in the way that tumor cells produce virus particles (*virions*). Virions are released from the cells by the budding process shown in Figure 1.9. A portion of the cell plasma membrane wraps around the maturing virus and pinches off. Millions of virions can be released without causing the death of the tumor cell. Rabbits injected with such cells synthesize antibodies that recognize both the foreign cells and the virions they produce.

The immunoprecipitates were formed by the following procedure. Normal or transformed cultured cells were lysed in a solution containing several detergents (such as SDS), and aggregates were removed by centrifugation at $200,000 \times g$ for 30 minutes. In step 1 of Figure 2.15, the supernatants, called the *cell extracts*, were incubated with 10 to 30 μl of antiserum (Ab) for 30 minutes at 4°C. This process resulted in the formation of complexes of cellular antigens (Ag) with antibodies (Ag-Ab). Then a small volume of *Staphylococcus aureus*, a bacterium containing protein A, was added (Figure 2.15, step 2). The A protein specifically binds to Ag-Ab complexes, causing the formation of large aggregates (*immunoprecipitates*) that quickly precipitate out of solution. These immunoprecipitates were washed several times and then prepared for the protein kinase activity assay.

The assay for protein kinase in the immunoprecipitates began with the resuspension of the bacterium-bound Ag-Ab complexes (Ag-Ab-A) directly in the enzyme reaction mixture (Figure 2.15, step 3). This mixture contained γ^{32}P-labeled ATP, which in the presence of protein kinase transfers ^{32}P to proteins present in the precipitate. After a brief incubation at 30°C, reactions were terminated by the addition of SDS, which also helped dissolve the aggregated proteins. After being heated at 95°C for one minute, the mixtures were centrifuged to remove aggregates, and the supernatants were analyzed either for acid-precipitable radioactivity (that is, ^{32}P-labeled protein) or by polyacrylamide gel electrophoresis (PAGE) autoradiography. Radioactivity in the gels was detected after the gels were dried on Whatman 3MM paper and then exposed to Kodak X-Omat R film. This film is sensitive to the β particles emitted by ^{32}P.

SDS-PAGE autoradiography of several normal and ASV-transformed chick cell cultures showed that protein phosphorylation occurred only in extracts of ASV-transformed cells that had been immunoprecipitated with serum from tumor-bearing rabbits; that is, TBR serum. Controls with normal serum did not give rise to radioactive immunoprecipitates. These results strongly support the conclusion that an ASV-coded protein in the transformed cells possesses protein kinase activity.

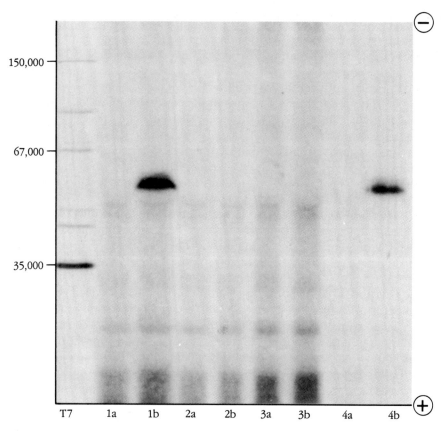

Figure 2.16 Presence of a protein kinase activity in immunoprecipitates of SR-ASV-transformed mammalian cells.
Cell extracts were prepared from cultures of SR-ASV-transformed baby hamster kidney cells (tracks 1), a normal phenotypic revertant clone of these cells (tracks 2), normal field vole cells (tracks 3), and SR-ASV-transformed field vole cells (tracks 4). Immunoprecipitation with either normal serum (a tracks) or serum from tumor-bearing rabbits (b tracks) and assay of phosphorylating activity were performed as outlined in Figure 2.15. The figure represents an autoradiogram of the dried gel. Phage T7 virion proteins were included in lane 1 as molecular weight markers.

Further evidence was needed to demonstrate that the protein phosphorylating activity observed in the extracts of ASV-infected chick cells was actually due to the viral *src* gene product. For this purpose, several mammalian cell lines were put through a similar analysis. The phosphorylating activity was detected in baby hamster kidney cells transformed by SR-ASV (Figure 2.16, track 1b). However, a normal *revertant* subclone of this cell line — that is, a mutant that no longer exhibited the transformed phenotype — did not display this enzymatic activity (track 2b). This line also did not contain p60^{src} in immunoprecipitates with TBR serum. A second mammalian cell type investigated was obtained from the European field mole. Again only the SR-ASV-transformed cell extracts that were immunoprecipitated with TBR serum contained the protein kinase activity (track 4b). Controls with immunoprecipitation by normal rabbit serum (tracks 1a, 2a, 3a, 4a) were all negative. Although these results strongly suggest that SR-ASV encodes a protein kinase, it is still not clear which viral gene is responsible for this activity.

2.4.3 Temperature Sensitivity of *src*-Encoded Protein Kinase

The use of a *conditional mutant* of the avian sarcoma virus — that is, a mutant virus that exhibits a particular phenotype under one set of conditions but not under another — provided nearly conclusive evidence that the *src* gene encodes the observed protein kinase activity. A temperature-sensitive conditional mutant of ASV, with the mutation mapping in the *src* gene, could transform cells at 35°C, but not at 41°C. Since nonmutated ASV can transform cells at 41°C and at 35°C, it is very likely that the protein encoded by the ts *src* gene was more heat sensitive than that of the ASV conditional mutant. Therefore, the *src* gene product is essential for cellular transformation. But what about protein kinase activity?

Cultures of chick cells infected with nondefective (nd) SR-ASV and with the temperature-sensitive *src* mutant NY68 were grown at both 30°C and 41°C. A cell extract was prepared from each of these four cultures, immunoprecipitated with TBR serum, and then analyzed as described above for protein kinase activity (Table 2.3). Whereas cells infected with the nondefective ASV had a somewhat higher phosphorylating activity at 41°C than at 35°C, the ts *src*-infected cells exhibited a greatly diminished activity (9 percent of normal) when they were grown at 41°C. Thus the protein kinase activity found in the immunoprecipitates of the ASV-infected chick cells was almost certainly encoded by the viral *src* gene.

Table 2.3. Temperature-sensitive (ts) expression of the *src* protein in chick cells infected with a ts transformation mutant (SR-NY68) of avian sarcoma virus (SR-ASV).

Virus	Growth temperature (°C)	Phosphorylating activity	
		^{32}P incorporated (fmol/mg protein)	Normalized values
SR-ASV	35	16.6	1.00
SR-ASV	41	35.8	2.16
SR-NY68	35	19.8	1.00
SR-NY68	41	1.7	0.09

2.4.4 Conclusions

The discovery of the protein kinase activity encoded by the *src* gene opened many avenues for research that has considerably enhanced our understanding of the molecular basis of cancer causation. It is now known that the *src* gene of avian sarcoma viruses is closely related to a family of genes found in all normal mammalian cells, the so-called *proto-oncogenes*. Several genetic mechanisms can convert these proto-oncogenes into *oncogenes*, which are responsible for the transformation of normal cells into tumorigenic cells. Many of the 40 characterized oncogenes code for protein kinase activities. The actual protein targets of these enzymes are being actively sought in many laboratories today because they could play key roles in the eventual prevention and cure of cancer.

2.5 The Relation of Protein Kinases to Evolution

The classical geneticist Theodosius Dobzhansky wrote that "nothing in biology makes sense except in the light of evolution." With the advent of genetic engineering technology (discussed in Chapter 5) and the desk-top computer, modern biologists have shed considerable light on the molecular process of evolution. A major portion of our current understanding has resulted from studies on the primary structure of proteins; that is, their amino acid sequences. Let us take a very brief look at this approach, using protein kinases once again as our example.

2.5.1 Gene and Protein Sequence Analysis

The conventional techniques for determining protein sequences by chemical analysis of purified proteins have been largely replaced in recent years by DNA sequencing methods, which are more easily performed once the gene that encodes the protein in question is in hand. One popular strategy for obtaining a specific gene proceeds as follows.

One first determines the sequence of a short stretch of amino acids (about 25) from the *amino terminal end* of the protein. The protein must be fairly pure for the sequence to be determined accurately. A still shorter segment—say, five to seven amino acids—within this peptide is then translated into a nucleotide sequence of DNA by means of the known genetic code. Since almost every amino acid is encoded by more than one codon, this step has some ambiguity. The investigator therefore must choose a peptide sequence that has a minimal degree of ambiguity and then synthesize all of the possible DNA molecules that can be translated into that peptide. These DNAs are made radioactive during their synthesis, and they serve as molecular probes to find the desired gene that contains their sequence. (The process is explained in Chapter 5.)

Once the gene has been located and cloned, its entire sequence can be determined rapidly. Using the genetic code once again, the investigator can deduce the precise amino acid sequence of the encoded protein. More than 5000 protein sequences are now known and available to anyone with a computer terminal. With this large data base, startling discoveries have already been made concerning the relation of protein structures to genes and evolution.

2.5.2 Protein Kinases Belong to a Gene Family

It is now recognized that a vast number of distinct but sequence-related protein kinases play a very large number of important roles. Besides being classified by the amino acid that serves as the phosphate acceptor, the protein kinase can also be categorized by the regulatory agent that modulates its enzymatic activity. In one scheme there are six groupings of this sort with regulatory agents including cyclic nucleotides, Ca^{++}, diacylglycerol, double-stranded RNA, hemoglobin, and the substrate. With such diversity in enzymes that catalyze the same chemical reaction, it is remarkable that the protein kinases can be assigned to a single gene family on the basis of their related amino acid sequences.

With the aid of a computer program, the sequences of *protein serine kinases* and *protein tyrosine kinases* are aligned as well as possible, with a 200-amino-acid segment of the catalytic subunit of bovine cAMP-dependent protein kinase serving as the reference peptide. Homologous amino acid sequences are evident in all of the protein kinases for which sequences are available. A best-fit alignment score has been assigned to each of them, so that an estimate of the percentage of sequence homology in relation to the reference peptide can be obtained. Some protein kinases, with their alignment scores in parentheses, include phosphorylase kinase (20.0), myosin light chain kinase (17.0), p60^{src} (12.0), and insulin receptor (9.0). The last is a protein tyrosine kinase that is autophosphorylated, and thereby activated, after it has bound to insulin. The extent of homology among these various proteins is sufficient to provide strong evidence that at least the catalytic subunits of many (all?) protein kinases are encoded by genes that originated from a common ancestral gene. This gene has apparently evolved into what is commonly referred to as a *gene family*. Many gene families are now known, and it is thought that they have arisen from gene duplication and/or amplification (see Chapter 1).

2.5.3   Oncogenes That Encode Protein Kinase Activity

At this time more than 40 oncogenes have been investigated. Their protein products can be placed in five categories on the basis of function, with the largest number having protein tyrosine kinase activity. Other groups include receptors that lack protein kinase activity, growth factors, and GTP binding proteins. Again striking homology was found within the DNA sequences of all the oncogenes that code for the catalytic subunits of protein kinases. This result provides strong evidence for the evolutionary relatedness of this group of oncogenes. This finding is particularly interesting when one realizes that related oncogenes appear in species as diverse as avian tumor viruses and normal mammalian cells, including our own.

Consider further the observation that a particular oncogene called *abl*, first seen in the Abelson leukemia virus, has now been found to be an essential gene in the fruit fly *D. melanogaster*. The *abl* gene encodes a protein with strong sequence similarity to a human cellular oncogene, *c-abl*. Moreover, when a portion of this gene was expressed in *E. coli*, it exhibited protein tyrosine kinase activity. These results suggest that the *abl* gene sequence has been conserved throughout evolutionary time, retaining homology during the evolution of both fruit flies and humans. The *abl* gene apparently encodes a protein tyrosine kinase, the function of which is essential to the development of the *Drosophila* pupa. Perhaps this and other oncogene products are also essential for the development of human embryos.

We can now more fully appreciate how protein purification and the determination of amino acid sequences bear directly on our understanding of areas as diverse as cancer, developmental biology, and evolution. It should be evident that the so-called classical fields of biology, as well as the newly emerging ones, require the understanding and application of modern technology. The general awareness and acceptance of this important principle helps the pace of biological research to accelerate at a rate that was unimaginable only a few years ago. For a quick demonstration of this point, one need merely plot the thickness of any common biological journal as a function of its year of publication.

Discussion Questions

1. Two proteins differ in net charge by one pK unit, but they have almost the same molecular weight. One is a protein kinase and the other is an ATPase. Of the three column chromatographic methods described (ion exchange, gel filtration, and affinity), which would be used to provide an effective separation of the two enzymes? Explain why each method is or is not appropriate.

2. A preparation of nuclei contained 10 mg of total protein and 200 units of protein kinase activity. After a high-salt precipitation step followed by ion-exchange chromatography, fractions numbered 30 to 32 contained 100 units of activity and 0.1 mg of protein.
   a. Calculate the specific activity of the protein kinase before and after purification.
   b. How much was the enzyme purified?

3. What might cause a partially purified enzyme fraction to appear to contain more than 100% of the enzymatic activity that was present in the original sample?

4. If an enzyme was known to possess four subunits of different but known molecular weights, what experiments would you perform to determine whether this enzyme was pure? Describe the expected results if it was indeed pure.

5. Increasing the concentration of cyclic AMP was shown to cause a marked stimulation of protein kinase activity until the level of 4×10^{-7} M was exceeded.
   a. Explain why a plateau value for enzyme activity is expected in an experiment of this type (Figure 2.13).
   b. If the plateau occurred at a fivefold greater concentration of cyclic AMP for a different preparation of this same protein kinase, what could you conclude about the relative purity of the two preparations? Assume that only protein kinase binds cyclic AMP in both preparations.

6. Describe an experiment by which you would determine the molecular weight of every polypeptide chain that served as a substrate for protein kinase in an immunoprecipitate that contained this enzymatic activity (Figures 2.15, 2.16).

7. What is the basis for concluding that protein kinases belong to a single gene family?

8. How did studies of protein kinases lead to conclusions about cancer and evolution?

References

Collett, M. S., and R. L. Erickson (1978). Protein kinase activity associated with the avian sarcoma virus *src* gene product. *Proceedings of the National Academy of Sciences USA, 75*:2021.

Doolittle, R. F. (1985). Proteins. *Scientific American, 252*:88.

Farron-Furstenthal, F. (1975). Protein kinases in hepatoma, and adult and fetal liver of the rat. I. Subcellular distribution. *Biochemical and Biophysical Research Communications, 67*:307.

Farron-Furstenthal, F., and J. R. Lightholder (1977). The purification of nuclear protein kinase by affinity chromatography. *FEBS Letters, 84*:313.

Hunter, T. (1984). The proteins of oncogenes. *Scientific American, 251*:70.

Kendrew, J. C. (1961). The three-dimensional structure of a protein molecule. *Scientific American, 205*:96.

Krebs, E. G. (1986). The enzymology of control by phosphorylation. In *The Enzymes* (P. D. Boyer and E. G. Krebs, eds.). Orlando, Florida: Academic Press.

Nishizuka, Y. (1984). Protein kinases in signal transduction. *Trends in Biochemical Science, 9*:163.

O'Farrell, P. H. (1975). High-resolution, two-dimensional electrophoresis of proteins. *Journal of Biological Chemistry, 250*:4007.

Taylor, S. S., J. Bubis, J. Toner-Webb, L. D. Saraswat, E. A. First, J. A. Buechler, D. R. Knighton, and J. Sowadski (1988). cAMP-dependent protein kinase: prototype for a family of enzymes. *FASEB Journal, 2*:2677.

Walsh, D. A., J. P. Perkins, and E. G. Krebs (1968). An adenosine 3',5'-monophosphate-dependent protein kinase from rabbit skeletal muscle. *Journal of Biological Chemistry, 243*:3763.

Walsh, K. A., H. Ericsson, D. C. Parmelee, and K. Titani (1981). Advances in protein sequencing. *Annual Review of Biochemistry, 50*:261.

3

The Discovery and Properties of Messenger RNA

3.1 Genes and Protein Synthesis

The central theme of molecular biology is generally recognized to be the molecular basis of gene function. Virtually all introductory biology courses teach that the coded information of the genetic material, DNA, is *transcribed* into *messenger RNA* (mRNA), which in turn is *translated* into protein (Figure 3.1). By means of the complex machinery of the cell, including structures such as ribosomes and the *endoplasmic reticulum*, the sequence of bases stored in DNA determines the sequence of amino acids in the entire array of cellular proteins.

Although this central dogma is well known by all students of biology, much is to be gained by a more detailed study of the discovery and properties of mRNA. As is often the case, advances in technology helped create a new branch of science — molecular biology in this instance. The major types of experimental procedures that opened the doors to molecular biology and became widely used in other areas of biological research include cesium chloride (CsCl) equilibrium density gradient centrifugation; sedimentation velocity analysis by centrifugation through sucrose density gradients; analysis of nucleic acid sequence homology by DNA-RNA hybridization; double-label counting of radioisotopes with a liquid scintillation spectrometer; and the pulse-chase procedure for determining the kinetics of synthesis and breakdown of macromolecules.

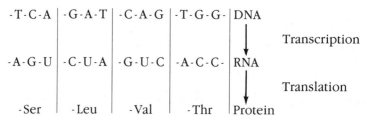

Figure 3.1 Information transfer from gene to protein.

The history of how RNA first came to be recognized as the intermediate between DNA and protein is not entirely clear. By the mid-1950s, after James Watson and Francis Crick discovered the structure of DNA, it was obvious to most scientists in the field that proteins were not translated directly from a DNA template. Ribosomes in prokaryotes and *microsomal particles* (clusters of ribosomes and membranes) in eukaryotes were shown to be the sites of protein synthesis. Because the microsomal particles were located in the cytoplasm, separated from DNA by a nuclear membrane, it seemed rather likely that some sort of "messenger" was required for the transfer of genetic information from the nucleus to the site of protein synthesis.

Even though bacteria have no nuclei, they became the "system" in which mRNA was discovered. Perhaps more surprising is the fact that this major breakthrough, together with many others in the new field of molecular biology, involved the collaboration of physicists and physical chemists with microbiologists, previously an unheard-of combination. Messenger RNA was discovered in 1961 in *E. coli* infected with the T2 *bacteriophage* (or *phage* for short), a virus whose host is a bacterial cell. This virus was unquestionably the most thoroughly studied at that time, primarily because the highly respected Caltech physicist Max Delbrück and a large number of his colleagues had been carrying out physicochemical, biochemical, and genetic experiments on T2 phage and its close relative T4 since the late 1940s.

An electron micrograph of T2 and an accompanying diagram of its complex morphological features are seen in Figure 3.2. Using its long proteinaceous tail fibers, the virus attaches itself with great specificity to a particular strain of *E. coli*. Shortly thereafter the sheath contracts, much like a muscle, and the viral DNA, stored within the icosahedral head structure, is injected into the bacterial host. The large number of biochemical events that take place during the next 20 minutes constitute the remainder of the life cycle of bacteriophage T2. Enzymes that the virus needs to replicate its DNA are synthesized, T2 DNA is replicated, viral coat proteins are synthesized, about 100 virus particles per cell are assembled, and the infected cells break open. Electron micrographs of

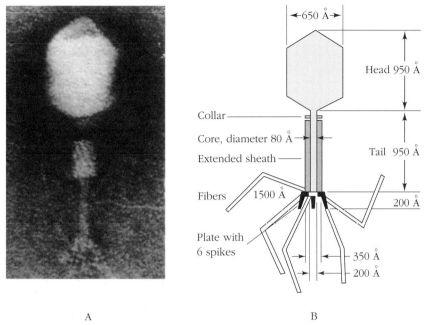

Figure 3.2   Detailed structure of bacteriophage T2.
(A) Electron micrograph. (B) Diagram.

ultrathin sections show the assembly of T2 viruses within infected *E. coli* at various stages of their development (Figure 3.3). The maturation of T2 viruses (virions) is preceded by the disappearance of the host cell's nucleoid region. This virus not only inhibits *E. coli* DNA replication but also degrades the bacterial genome down to its constituent deoxyribonucleotides. As we shall see, the virulence of T2 was of considerable benefit in the discovery of messenger RNA.

3.2   The Discovery of T2 Messenger RNA

Two elegant papers published in 1961 conclusively demonstrated that there existed, at least for bacteriophage T2 and its cousin T4, an mRNA intermediate in the transfer of genetic information from DNA to protein. The research reported in both papers used new technologies that became widely applicable in the embryonic field of molecular biology: equilibrium density gradient centrifugation and DNA-RNA hybridization analysis.

   In May 1961 a paper titled "An Unstable Intermediate Carrying Information from Genes to Ribosomes for Protein Synthesis" appeared

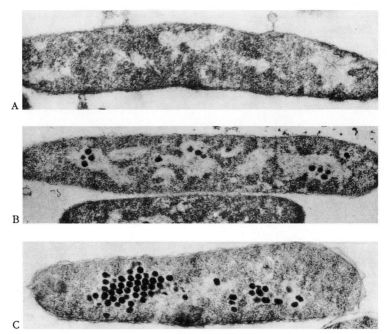

Figure 3.3 Electron micrographs of ultrathin sections of T2-infected *E. coli* at various stages of intracellular phage growth.
(A) Ten minutes after infection; the bacterial nucleoid regions have disappeared and vacuoles filled with fibrillar material (phage DNA) have appeared. (B) Fourteen minutes after infection; the first condensates of phage DNA have been formed. (C) Forty minutes after infection; many condensates and structurally intact phage heads are present.

in *Nature*. It described experiments conducted at the California Institute of Technology by three well-known molecular biologists: Sidney Brenner, of the University of Cambridge, England; François Jacob, of the Institut Pasteur in Paris; and Matthew Meselson, of Caltech. Meselson, together with Franklin Stahl, had earlier developed the CsCl equilibrium density gradient procedure in showing that DNA replicates by a semi-conservative mechanism. That is, each parental strand of the double-helical DNA molecule becomes associated with a newly synthesized complementary strand, following the base pairing rules of Watson and Crick. Now Brenner and his colleagues used the same procedure for yet another classic discovery.

3.2.1 Equilibrium Density Gradient Centrifugation

Before we get into the details of the experimental design used by the team at Caltech, it is important first to understand the principles behind the equilibrium density gradient technique (Figure 3.4). In this

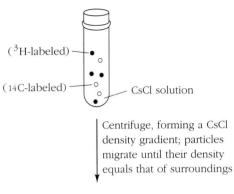

Molecules or particles of different
density and radioactive label are
mixed with CsCl solution

 • ^{3}H-labeled particles/molecules
 ○ ^{14}C-labeled particles/molecules

(^{3}H-labeled)

(^{14}C-labeled)

CsCl solution

Centrifuge, forming a CsCl
density gradient; particles
migrate until their density
equals that of surroundings

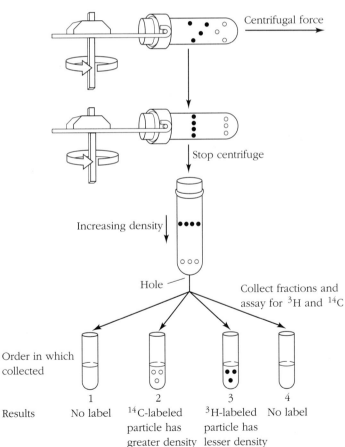

Centrifugal force

Stop centrifuge

Increasing density

Hole

Collect fractions and
assay for ^{3}H and ^{14}C

Order in which
collected

1	2	3	4
Results	No label	^{14}C-labeled particle has greater density	^{3}H-labeled particle has lesser density	No label

Figure 3.4 Equilibrium density gradient centrifugation.

procedure, a solution containing a high-density solute such as CsCl is subjected to a centrifugal force in an ultracentrifuge. Present in the solution is a sample of macromolecules (such as DNA, RNA, or protein) or particles (such as ribosomes) that is of interest to the investigator. The sample chosen by Brenner and his associates consisted of *ribosomes*, particles containing nearly equal weights of RNA and protein. During centrifugation at high centrifugal forces, the CsCl rapidly (within two to three hours) forms a linear density gradient throughout the centrifuge tube, with the highest density at the bottom, farthest from the center of rotation. The tube is held in a swing-out "bucket" that is attached to the rotor, generally referred to as a *swinging bucket rotor* (see Figure 3.4). The macromolecules and particles in the CsCl solution sediment or float under the centrifugal force, depending on their buoyant density, and eventually (after 48 to 72 hours) reach equilibrium. This position corresponds to the level at which their buoyant density equals that of the CsCl solution. Further centrifugation would not change the position of the macromolecules in the gradient.

After equilibrium is reached, centrifugation is stopped and samples (*fractions*) are collected through a hole punctured in the bottom of the tube. Recall that density is equal to mass per unit of volume. Thus molecules of differing molecular weights, such as DNA fragments, may have the same density. Alternatively, molecules that have nearly identical molecular weights but occupy different volumes might have sufficiently different densities in a CsCl gradient to permit their separation. In fact, the number of cesium ions bound to a molecule is more important than the molecule's volume. For example, more Cs^+ binds to RNA than to DNA per unit of mass, so that RNA has greater density in a CsCl gradient.

3.2.2   Association of T4 mRNA with *E. coli* Ribosomes

Brenner and his colleagues designed their experiment to distinguish among three alternative theoretical models by which genes might direct the synthesis of proteins. It had already been established that T4 phage, although it is composed solely of DNA and protein, directs the synthesis of a new type of RNA in the infected bacteria. This RNA has a base composition much closer to that of T4 DNA than to that of *E. coli* DNA. Model 1 proposed that T4 virus directed the synthesis of a new class of ribosomes that possessed the necessary genetic information for the synthesis of T4 proteins. Model 2 proposed that

> A special type of RNA molecule, or "messenger RNA," exists which brings genetic information from genes to non-specialized ribosomes, and that the consequences of phage infection are two-fold; a) to switch off the synthesis of new ribosomes; b) to substitute phage

messenger RNA for bacterial messenger RNA. (Brenner, Jacob, and Meselson, 1961, p. 576)

For simplicity we can skip discussion of the third model, in which DNA is copied directly into viral proteins and the viral-directed RNA prevents ribosomes from making bacterial proteins.

As in the already famous experiment of Meselson and Stahl, the Brenner team employed the procedure of density-labeling the cells. One culture medium contained ^{15}N (99 percent) and ^{13}C (60 percent) as high-density isotopic labels and $^{32}PO_4$ as a radioactive label for the cell's nucleic acids. A 50-fold greater quantity of a second culture was also prepared and mixed with the first. This second culture served as a *marker* preparation: its ribosomes were detected in the CsCl gradient fractions by virtue of their absorbance of ultraviolet (UV) light; that is, *optical density* at 254 nm (OD_{254}). A *spectrophotometer* is used for the accurate measurement of OD. The large excess of the unlabeled culture enables it to obscure the OD of the ^{32}P-labeled heavy ribosomes. By this means, ribosomes from both cultures can be detected in the same CsCl gradient, one by virtue of UV absorbance and the other by radioactivity. Thus even very slight differences in buoyant density of the ribosomes from the two cultures are significant in such an experiment.

The density profiles of ribosomes purified from heavy-labeled *E. coli* and from cells grown in normal light medium are shown in Figure 3.5. Note that in both cases, two peaks of ribosomes were present in the gradients. The reason for the two peaks is unclear. The authors' remarks about their distribution as a function of Mg^{++} concentration seem to indicate that the high-density band was composed of 50S and 30S ribosomes and the low-density band contained the 70S ribosomes (the meaning of S-values will become clear in Section 3.3). Other researchers had shown earlier that 70S ribosomes are active in protein synthesis. Each 70S ribosome is assembled from the combination of one 30S and one 50S ribosome. This control experiment, in which heavy and light ribosomes were mixed before they were centrifuged in CsCl, demonstrated that there was no spurious aggregation between the ribosomes of different isotopic compositions.

The experimenters then performed a similar experiment with T4-infected *E. coli*. As Figure 3.6 shows, a large fraction of the ^{14}C-labeled RNA that was synthesized from the third to the fifth minute after infection banded with a density in the position of the 70S ribosomes (low-density peak). Note that there is no peak of ^{14}C-labeled RNA in the position of the higher density ribosome peak—that is, the 30S and 50S ribosomes. Therefore, not all cellular ribosomes were labeled equally after infection. Moreover, a substantial portion of the pulse-labeled ^{14}C-RNA was found near the bottom of the gradient, which is a characteristic of *free* RNA (RNA not associated with ribosomes). These results indicate that after infection of *E. coli* by T4 phage, a new type of RNA and/or

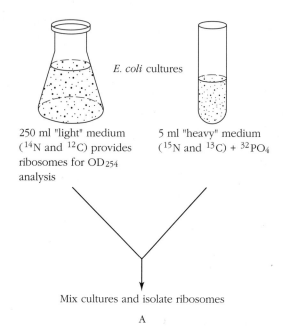

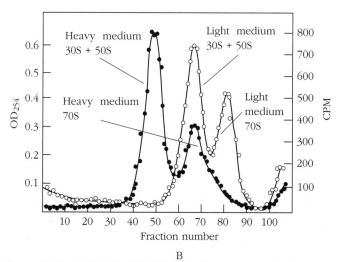

Figure 3.5 Control experiment of Brenner, Jacob, and Meselson.
(A) Protocol for experiment. (B) Distribution of heavy and light ribosomes in a CsCl density gradient. *E. coli* were grown in 5 ml of a heavy medium containing ^{15}N (99%) and ^{13}C (60%) algal hydrolysate and radioactive phosphate ($^{32}PO_4$). These cells were mixed with a 50-fold excess of cells grown in the usual light medium. Ribosomes were extracted, purified by centrifugation, and then centrifuged to their position of equilibrium density in CsCl (35 hours at 37,000 rpm). Ultraviolet (UV) absorption at 254 nm detects the excess light ribosomes (o), and ^{32}P counts detect the heavy ribosomes (•).

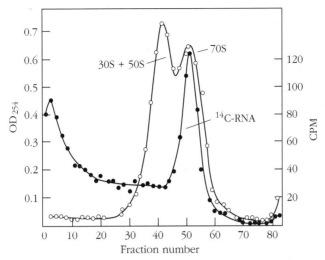

Figure 3.6 Density distribution of ribosomes and of RNA synthesized after T4 phage infection.
A culture of T4-infected *E. coli* was labeled with ^{14}C-uracil from the third to the fifth minute after infection. Purified ribosomes were prepared and centrifuged to equilibrium in CsCl. Fractions were collected and analyzed for UV absorption (o) and ^{14}C-RNA (•).

ribosomes is synthesized. Two possibilities exist. Either the ^{14}C-RNA found in the position of 70S ribosomes is a component of some new type of ribosome synthesized after T4 infection, or this ^{14}C-RNA is associated with preformed *E. coli* 70S ribosomes. If the latter were true, it would be evidence of a property expected of viral-coded messenger RNA. If the ^{14}C-RNA is actually a component of a new class of 70S ribosomes, then these ribosomes must be unusually fragile in CsCl, judging from the ^{14}C profile in the gradient.

To distinguish between these two possibilities (models 1 and 2), the investigators performed an isotope transfer experiment in accordance with Meselson and Stahl's procedure. Cells grown in a small volume of "heavy" medium ([15]N[13]C) were infected with T4 phage, transferred to "light" medium ([14]N[12]C), and radiolabeled with [32]PO₄ from the second to the seventh minute after infection. These cells were mixed with a 50-fold excess of marker cells grown and infected in a light nonradioactive medium. As we have seen, the marker cells provide the ribosomes that can be detected by optical density (OD$_{254}$). The ^{32}P-labeled RNA, synthesized in the heavy medium, banded in the position of the marker 30S and 50S ribosomes of the cells grown in light medium (Figure 3.7). But recall fom the earlier experiment (Figure 3.6) that the RNA synthesized after infection banded in the region of the 70S ribosomes, not in the position of the 30S and 50S ribosomes. Brenner and his colleagues therefore concluded that the viral-directed ^{32}P-RNA became

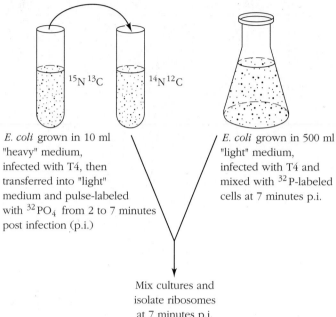

E. coli grown in 10 ml
"heavy" medium,
infected with T4, then
transferred into "light"
medium and pulse-labeled
with $^{32}PO_4$ from 2 to 7 minutes
post infection (p.i.)

E. coli grown in 500 ml
"light" medium,
infected with T4 and
mixed with ^{32}P-labeled
cells at 7 minutes p.i.

Mix cultures and
isolate ribosomes
at 7 minutes p.i.

A

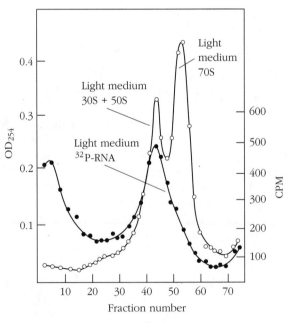

B

associated with the heavy 70S ribosomes that were constructed before infection, while the cells were growing in the heavy medium.

This ingenious set of experiments demonstrated rather convincingly that the bacterial virus T4 synthesizes a messenger RNA that associates with preexisting 70S ribosomes of the infected host cell, *E. coli*. Model 2 was proved to be correct. T4 phage was a fortunate choice because it shuts off the synthesis of all types of host RNA. Most viruses do not exhibit such a strongly virulent behavior. With phages T2 and T4, only viral RNA becomes labeled after infection, so it was possible to distinguish it from cellular RNA by the labeling procedure employed. These experiments also led to the important conclusion that "ribosomes are non-specialized structures which synthesize, at a given time, the protein dictated by the messenger they happen to contain" (Brenner, Jacob, and Meselson, 1961, p. 576). It was not long before similar conclusions were reached with uninfected prokaryotes and eukaryotes.

3.2.3   Sequence Homology Between T2 Messenger RNA and T2 DNA

An equally ingenious set of experiments, though much simpler to understand, also provided convincing evidence that T2 phage directs the synthesis of a new class of RNA after infection of *E. coli*. In these experiments, carried out at the University of Illinois, Benjamin Hall, a physical chemist, and Sol Spiegelman, a microbiologist, used their own recently developed technique referred to as *DNA-RNA hybridization*. By this method it immediately became apparent that the virus-directed RNA possesses a sequence that is *homologous (complementary)* to that of the viral DNA. The transfer of genetic information from DNA to this messenger RNA obviously occurs via the same base pairing rules that permit DNA to duplicate itself faithfully.

The experimental design was based on a recent exciting discovery made at Harvard by Hall's Ph.D. thesis adviser, Paul Doty, and his student Julius Marmur. The Harvard physical chemists reported at a

---

Figure 3.7   Cell transfer experiment of Brenner, Jacob, and Meselson. (A) Protocol for experiment. (B) Density of RNA synthesized by T4-infected cells after transfer from heavy to light medium. *E. coli* were grown in 10 ml of heavy (^{15}N, ^{13}C) medium, infected with T4 phage, and then switched to a light (^{14}N, ^{12}C) medium. The cells were radiolabeled with ^{32}PO$_4$ from the second to the seventh minute after infection in the light medium. This culture was mixed with a 50-fold excess of *E. coli* grown in light medium. Purified ribosomes were prepared and centrifuged to equilibrium in CsCl. Fractions were collected and analyzed for UV absorption (o), which detects the light ribosomes, and for ^{32}P (●), which detects the RNA synthesized in the heavy cells that were transferred after infection to a light medium.

convention in the summer of 1960 that the complementary strands of heat-denatured (single-stranded) DNA could reconstitute the native double-helical DNA structure when they were cooled extremely slowly. This *renaturation* process occurred only between DNA strands that originated in the same or closely related organisms, implying that the specificity for this physicochemical reaction resided in the sequence complementarity of the bases in the two DNA strands.

Ignoring the controversy generated by the famous iconoclast Max Delbrück over the accuracy of this report, Hall and Spiegelman quickly adapted Marmur and Doty's experimental procedure to an analysis of the RNA synthesized after infection of *E. coli* by T2 phage. They reasoned that if T2 directed the synthesis of a new type of RNA that contained the genetic information for the synthesis of viral proteins, then this T2-specific RNA should possess a nucleotide sequence that is homologous to some portion of the T2 DNA molecule. Denatured T2 DNA should then be capable of forming a DNA-RNA hybrid with T2-specific RNA when the mixture was slowly cooled.

Hall and Spiegelman infected a culture of *E. coli* with T2 phage and then radiolabeled it with $^{32}PO_4$ for three to eight minutes. They purified ^{32}P-RNA from the infected cells, mixed it with denatured T2 DNA, and subjected the mixture to slow cooling from 65°C to room temperature. For this purpose they used an insulated water bath with a 40-liter capacity. The "magical" slow-cooling process required 30 hours. The experimenters analyzed DNA-RNA hybrid formation by means of CsCl equilibrium density gradient centrifugation. This procedure required centrifugation for five days at 33,000 rpm. Hall and Spiegelman then collected fractions and analyzed them for radioactivity and optical density at a wavelength of 260 nm. These experiments also made use of the newly developed procedure of *double-labeling* analysis with a liquid scintillation spectrometer. With this instrument, two radioisotopes such as ^{32}P and ^{3}H that are present in the same sample can be counted separately (see Chapter 2).

The striking results made this tedious procedure very worthwhile. The CsCl gradient completely separated ^{32}P-RNA from heat-denatured, ^{3}H-labeled T2 DNA in mixtures that were not subjected to slow cooling (Figure 3.8). In fact, denatured T2 ^{3}H-DNA was even separated from native T2 DNA, which was analyzed in the gradient fractions by OD. Most exciting was the next gradient, in which a peak of ^{32}P-RNA appeared in the region of the denatured ^{3}H-DNA as a result of the slow cooling of the mixture (Figure 3.9). Further experiments showed that the presence of the ^{32}P-RNA at the density of denatured DNA required the presence of denatured T2 DNA during the slow-cooling procedure. Additional controls indicated that when denatured DNA was obtained from *E. coli* and other organisms unrelated to bacteriophage T2 and mixed with the ^{32}P-RNA extracted from T2-infected cells, no DNA-RNA hybrid was formed. Hall and Spiegelman therefore concluded that the

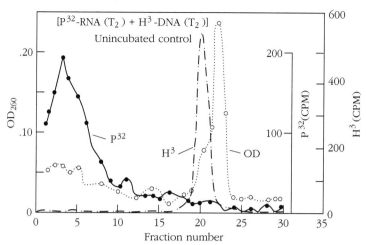

Figure 3.8   Separation of ^{32}P-RNA from T2 ^{3}H-DNA by CsCl density gradient centrifugation.
A mixture of heat-denatured T2 ^{3}H-DNA, ^{32}P-RNA synthesized from the third to the eighth minute after T2 infection of *E. coli*, and unlabeled native T2 DNA was centrifuged to equilibrium in CsCl (5 days at 33,000 rpm).

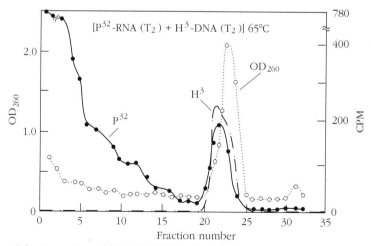

Figure 3.9   Detection of T2 DNA-RNA hybrid formation.
^{32}P-RNA from T2-infected *E. coli* and heat-denatured T2 ^{3}H-DNA were mixed and slowly cooled from 65°C to 26°C. Unlabeled native T2 DNA and CsCl were then added and the mixture was centrifuged to equilibrium.

^{32}P-RNA synthesized after T2 infection had sequences homologous to T2 DNA, and that their DNA-RNA hybridization procedure possessed the specificity needed for the detection of this homology. Yet another technique with widespread application was born, and by the spring of 1961 the concept of messenger RNA was firmly established.

3.3 Properties of Prokaryotic Messenger RNA

3.3.1 Sucrose Density Gradient Centrifugation

Many types of macromolecules and subcellular particles are characterized by, and even named for, their sedimentation coefficient, or *S-value*. For example, bacteria possess 30S, 50S, and 70S ribosomes, which contain 16S and 23S ribosomal RNA (rRNA). The S-value is directly proportional to a particle's mass (m) and indirectly to its frictional coefficient (f):

$$S = km/f$$

This helps explain why one 30S ribosome combines with one 50S ribosome to produce a 70S ribosome, not an 80S particle. While the 30S and 50S ribosomes are spherical, the 70S particle is dumbbell-shaped and has a greater frictional coefficient during sedimentation.

In the early 1960s the *sucrose density gradient centrifugation* procedure was developed for the separation of macromolecules. This technique took advantage of differences in S-values between different classes of RNA, for example, and was used for both preparative and analytical purposes. Because molecules of different sizes were found in different zones of the centrifuge tube, the technique was also referred to as *rate-zonal centrifugation*. It should be emphasized that this technique separates molecules on the basis of their mass, not their density. The purpose of the sucrose density gradient is to retard the rate of diffusion of the separated macromolecules during and after centrifugation.

Figure 3.10 shows how the technique is carried out. The investigator prepares a preformed linear sucrose gradient directly in the centrifuge tube, using a mixing chamber and two solutions of sucrose; for example, 5 percent and 20 percent sucrose. The sample solution, having a density less than that of 5 percent sucrose, is layered on top of the 5 to 20 percent sucrose density gradient. The gradients are then placed into a swinging bucket rotor and centrifuged at very high gravitational forces (often in excess of $100,000 \times$ g) in the ultracentrifuge. The samples are spun long enough to achieve the desired separation. If they are spun too long, all molecules will sediment to the bottom of the tube. Note the difference between this procedure and equilibrium density gradient

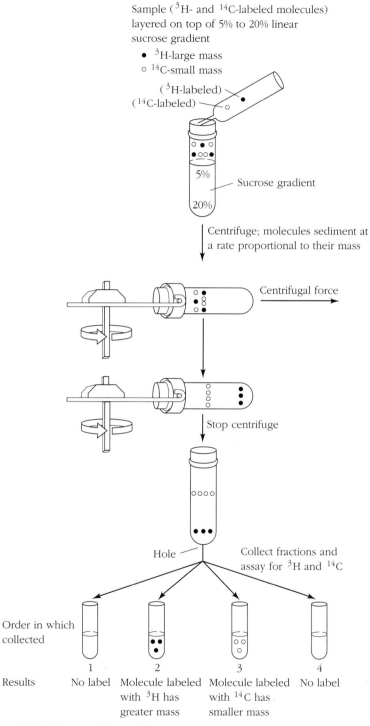

Figure 3.10 Rate-zonal centrifugation.

centrifugation, in which a molecule reaches its position of equilibrium density after a certain centrifugation time and then remains there indefinitely (see Figure 3.4).

3.3.2   Size Heterogeneity of T2 mRNA

The existence of a new class of RNA synthesized after T2 infection of *E. coli* is also readily apparent upon examination of the data obtained from a sucrose density gradient sedimentation experiment (Figure 3.11). RNA was obtained from a culture of bacteria labeled with ^{3}H-uridine from five to six minutes after infection with T2 phage. Taking precautions to avoid breakdown of the RNA during the extraction procedure, Bernard Sagik and his coworkers in Sol Spiegelman's laboratory noted the extremely heterogeneous sedimentation rate profile of the ^{3}H-RNA synthesized after infection. The OD_{260} profile in the gradient displayed the 23S and 16S *E. coli* ribosomal RNA. The ^{3}H-RNA in the same gradient, by contrast, had S-values ranging from 4S to 25S. The DNA-RNA hybrid test developed by Hall and Spiegelman showed that all size regions contained T2 mRNA.

The extensive heterogeneity in the size of T2-specific RNA is an expected property of mRNA. As we know, proteins range in molecular weights from about 5000 to well over 100,000 daltons. With a coding

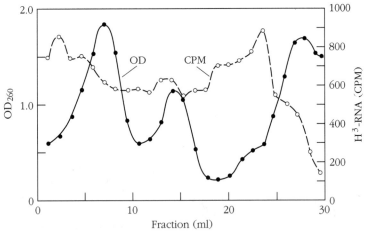

Figure 3.11   Sedimentation analysis of RNA.
Separate aliquots of T2-infected *E. coli* were pulse-labeled with H^3-uridine from the fifth to the sixth minute after infection. RNA, isolated by phenol extraction of whole cell extracts, was immediately subjected to sucrose density gradient centrifugation for 11.5 hours at 25,000 rpm. Fractions were collected from the bottom of the centrifuge tube (0 ml) and analyzed for UV absorbance at a wavelength of 260 nm (OD_{260}) and for acid-precipitable H^3-RNA (CPM).

ratio of 3 nucleotides per amino acid, the corresponding mRNAs would range in molecular weight from about 50,000 to more than 1,000,000. Such RNAs would have the heterogeneous S-values observed by Sagik and his colleagues. In sucrose gradients, the mass of spherical molecules such as randomly coiled RNA is directly proportional to the distance traveled through the gradient. Thus molecular weights of the ^{3}H-RNA molecules can be estimated from the positions of the 16S (500,000 d) and 23S (1,000,000 d) rRNAs in the gradient.

3.3.3 The Instability of T2 mRNA

The commercial availability of radioactive metabolites led to another very fundamental and ingenious experimental technique, the *pulse-chase experiment.* During the "pulse," cells are incubated for a brief period with a radiolabeled metabolite such as $^{32}PO_4$. This process serves to label newly synthesized molecules such as RNA and DNA. In the subsequent "chase," the radioactive compound in the growth medium is removed, generally by centrifuging or filtering the cells, and the cells are then grown in a medium containing a large amount of nonradioactive ("cold") metabolite ($^{31}PO_4$). The cold metabolite is rapidly transported into the cells, and it dilutes the specific activity of the residual intracellular pool of $^{32}PO_4$. By this means, the phosphorus-containing molecules synthesized during the chase period become much less radioactive than those synthesized during the pulse-labeling step.

It thus becomes possible during the chase to follow the fate of the molecules that were radiolabeled during the pulse. In practice, this procedure is generally employed to determine the relative stability of the labeled molecules or to follow their change in location from the time of synthesis. Because macromolecules are acid precipitable, whereas small molecules such as the radioactive precursors are not, the assay for macromolecular degradation ("turnover") is quite simple. One must be careful, however, to remove all of the radioactive precursor, for it is generally present far in excess of the newly synthesized macromolecules.

Using the pulse-chase procedure, Lazarus Astrachan and Elliot Volkin at the Oak Ridge National Laboratory in Tennessee analyzed the kinetic properties of DNA and RNA synthesized in *E. coli* after infection by T2 phage. They also used the antibiotic *chloramphenicol* (CP) as a means of providing further useful information. This drug completely inhibits protein synthesis in bacteria. The results shown in Figures 3.12A and B were obtained with CP added one and nine minutes after infection, respectively. In both control experiments ("Con"), it was evident that most of the acid-precipitable ^{32}P-labeled RNA synthesized during the pulse in $^{32}PO_4$ was rapidly degraded during the chase. Thus primarily unstable RNA was synthesized during the pulse with $^{32}PO_4$. When CP

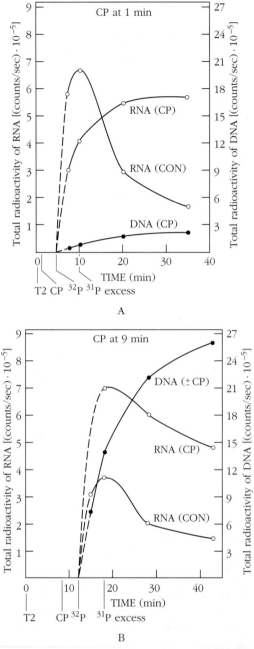

Figure 3.12 Turnover of ^{32}P-RNA and accumulation of ^{32}P-DNA as a function of time after T2 phage infection: effects of chloramphenicol (CP). (A) CP added at 1 minute. (B) CP added at 9 minutes. To parallel samples of infected cells, ^{32}P-labeled phosphate was added at designated times after infection. Six minutes after ^{32}P was added, its specific radioactivity was reduced by adding sufficient phosphate buffer (^{31}P) to increase the phosphate concentration from 1.5 μg of P/ml to 225 μg of P/ml. When chloramphenicol (CP) was used, it was added 3 minutes before ^{32}P. Control (CON) indicates no chloramphenicol added.

was added, however, there was considerably less degradation of the newly synthesized ^{32}P-RNA. This effect was most pronounced when CP was added very soon (one minute) after infection. At this early time, but not at nine minutes, CP also inhibited the synthesis of DNA.

Although these findings were not thoroughly understood in 1959, when this paper was published, later work has provided a clear interpretation. First, and of most relevance, it may be concluded from the control experiments that T2 mRNA has a relatively short half-life, on the order of only three minutes. This is a unique property of mRNA, distinguishing it from ribosomal and transfer RNAs. Second, we now know that T2 phage directs the formation of proteins that quickly and completely shut down the synthesis of all types of *E. coli* RNA, including stable ribosomal and transfer RNA. This inhibitory effect of T2 is prevented by CP if it is added before the production of T2-coded proteins. Therefore, the ^{32}P-RNA synthesized in the presence of CP was primarily stable *E. coli* RNA instead of unstable T2 mRNA, especially when the drug was added very shortly after infection. Note that the final plateau attained during the chase indicates the proportion of stable RNA that was synthesized during the pulse-labeling step. Finally, because T2-coded proteins are required for viral DNA synthesis, CP added shortly after infection would also be expected to reduce the amount of ^{32}P-DNA formed.

It is evident that the pulse-chase procedure can provide some very basic information about the synthesis, degradation, and localization of viral and cellular components. The utility of this technique cannot be overestimated. The use of metabolic inhibitors, as in these experiments, also has widespread applications in research and chemotherapy.

3.4   Eukaryotic Messenger RNA

Just when you think that everything important has been learned about a particular subject, new discoveries lead to unimagined concepts. Consider the research on eukaryotic mRNA. It is doubtful whether anyone guessed that the characterization of this mRNA would provide startling new concepts that were not foretold by the results obtained with prokaryotic mRNA. Let us take a look at some of the experiments that revealed the most remarkable properties of eukaryotic mRNA.

3.4.1   Processing Heterogeneous Nuclear RNA (hnRNA)

After mRNA was discovered in bacteria, it was obvious that eukaryotes must solve a logistics problem that did not exist in prokaryotes; namely, the transfer of mRNA from its nuclear site of synthesis to its cytoplasmic site of function. The same problem was posed for the other types of RNA as well, since it had been established earlier that ribosomes and transfer RNAs reside in the cytoplasm. Pulse-chase experiments indicated that newly synthesized RNA, made radioactive by the incorporation of ^{3}H-uridine during a short pulse of about 10 minutes, was all located initially in the nucleus. During the chase period, however, when the cells were transferred to nonradiolabeled medium, only a small fraction (10 to 20 percent) of the nuclear ^{3}H-RNA was transported to the cytoplasm. The remainder was ultimately broken down to nucleotides (AMP, GMP, CMP, and UMP—see Appendix 2 for definitions). Such experiments, which used both autoradiography and cell fractionation procedures, yielded the surprising hypothesis that eukaryotic cells are quite wasteful of energy, uneconomically degrading 80 to 90 percent of the total RNA synthesized.

After this puzzling finding became generally recognized as valid, the barrage of discoveries concerning RNA processing steps became somewhat easier to accept, if not to understand. Although all types of RNA have been shown to undergo a variety of molecular processing events, we shall focus only on some that affect mRNA. One of the most direct ways of demonstrating RNA processing entails sedimentation rate analysis. In order to compare the S-values of nuclear RNA with those of mRNA, it was necessary to separate mRNA from the other types of cytoplasmic RNAs; that is, ribosomal and transfer RNAs. This process became feasible around 1970, when it was discovered that most mRNA molecules and a large fraction of nuclear RNAs possess a polyadenylic acid (poly A) segment, 150 to 200 nucleotides long, at their 3' *terminus.* The poly A segment was soon shown to be added as a posttranscriptional nuclear processing event that occurs specifically on RNA molecules destined to serve as messengers for protein synthesis. A variety of techniques was rapidly developed to exploit this property for purifying mRNA away from the other types of cellular RNA.

A comparison of the size distribution of mRNA and nuclear RNA extracted from mouse L cells, a mammalian cell line that grows indefinitely in culture, is shown in Figure 3.13. Nuclear RNA, pulse-labeled for 25 minutes with ^{14}C-uridine, was purified under conditions that minimized degradation. The mRNA was labeled for four hours with ^{3}H-uridine and isolated from ribosomes under conditions that selected for RNA containing poly A. To obtain a valid estimate of the molecular size distribution, the nuclear RNA and mRNA samples were subjected to strong denaturation conditions, so that all secondary structure was destroyed and all molecules were converted to the random coil configuration. In this process the RNA was heated to 60°C in 80 percent dimethylsulfoxide or a similar organic solvent.

Nuclear RNA displays a broad range of sizes, from about 0.5 to 30 kilobases (kb), with an average mass of about 13 kb (Figure 3.13A). For this reason, such RNA has been called *heterogeneous nuclear RNA,* or *hnRNA.* The mRNA is much smaller, ranging from about 0.5 to 20 kb,

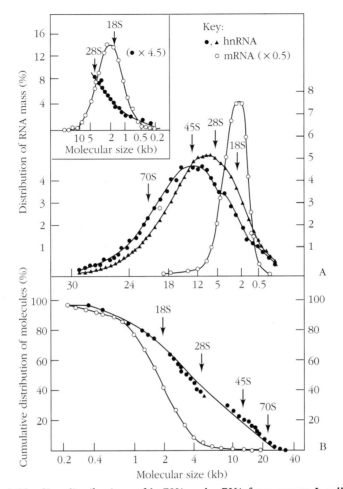

Figure 3.13   Size distributions of hnRNA and mRNA from mouse L cells.
(A) Two different preparations of hnRNA (● and ▲) from cells labeled for 25
minutes with ^{14}C-uridine were denatured and sedimented through 15% to 30%
sucrose gradients containing 0.5% sodium dodecyl sulfate. The gradients were
fractioned into 44 samples, and the CPM in each sample were assayed and
divided by the total CPM on the gradient to give percent distribution of RNA
mass. Polyadenylated mRNA from polyribosomes of cells labeled for 4 hours
with ^{3}H-uridine (○) was denatured by heat and chemicals and sedimented
with ^{14}C-labeled rRNA markers. The abscissas were converted from fraction
number to molecular size in kilobases (kb) by means of calibration curves
constructed with 18S and 28S rRNA, 45S pre-rRNA, and 70S Rous sarcoma virus
RNA, assumed to be 2, 5, 12.5, and 22 kb, respectively. (B) The mass
distributions shown in (A) were converted to molecular distributions by
dividing each ordinate value by its abscissa value. The data were plotted
cumulatively as percent molecules greater than a particular size. A separate
gradient was run for the hnRNA sedimenting more slowly than 50S so that
better resolution could be achieved (see *inset*). The symbol (△) represents
the point at which data for the gradient of <50S hnRNA were normalized to
data for large hnRNA.

with an average mass of about 3 kb. The approximate numerical weight distribution of these RNAs in Figure 3.13A was obtained by dividing each ordinate (y-axis) value by its abscissa (x-axis) value. After these results were plotted cumulatively as percent molecules greater than a particular size (Figure 3.13B), a clearer indication of the relative number of RNA molecules of each size class became apparent. It is evident from such experiments that the great majority of the nuclear RNA molecules are reduced in size before they become functional mRNA molecules in the cytoplasm.

3.4.2 Detection and Function of Polyribosomes

As in prokaryotes, mRNA in eukaryotes is less stable than ribosomal and transfer RNAs. Instability confers the advantage of enabling a cell to shift gears, so to speak, and synthesize a new set of proteins fairly quickly when necessary. For example, each type of differentiated cell synthesizes a highly specialized set of proteins in addition to those needed for the maintenance of life. Also, as cells pass through different stages of their growth cycle (mitosis, G_1, S, and G_2), the synthesis of certain proteins must be turned on or off fairly rapidly.

The efficiency of mRNA in protein synthesis is enhanced in the cell by the formation of *polyribosomes (polysomes)*. Polysomes consist of one mRNA associated with multiple ribosomes. This structure enables the mRNA to be translated into several copies of its encoded protein simultaneously. One of the first demonstrations of the existence of polysomes occurred in 1963 in the laboratory of Alexander Rich. Anticipating that the structure of polysomes would be quite fragile, Rich and his coworkers used rabbit reticulocytes as a starting material. *Reticulocytes*, which are precursors to mature red blood cells, can be broken open with a minimum of mechanical manipulation and shearing forces.

Blood containing 80 to 90 percent reticulocytes was collected by puncturing the hearts of rabbits made anemic with phenylhydrazine. The cells were washed and then pulse-labeled for 15 minutes at 37°C with a mixture of ^{14}C-labeled amino acids or ^{3}H-leucine. The cells were washed again by centrifugation and gently lysed by osmotic shock in a hypotonic solution. Cell membranes were removed by centrifugation at 10,000 × g for 15 minutes, and the supernatant was carefully decanted. Sedimentation through sucrose gradients revealed the existence of rapidly sedimenting material that was highly active in protein synthesis (Figure 3.14). This material sedimented considerably faster than the 78S ribosome previously thought to be responsible for protein synthesis. Using sedimentation markers, the rapidly sedimenting material with labeled *nascent* (newly synthesized) protein was shown to have S-values ranging from about 100S to 170S (Figure 3.14).

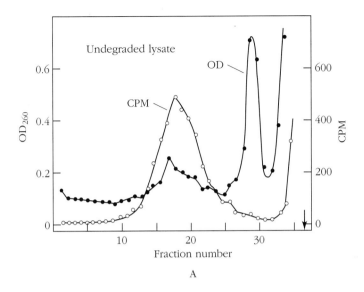

A

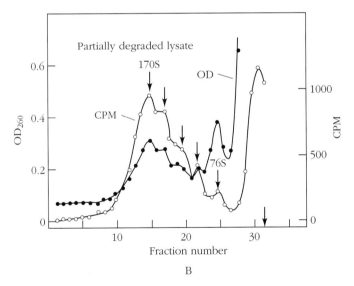

B

Figure 3.14   The association of newly synthesized protein with polyribosomes in reticulocytes.
After a brief incubation of reticulocytes with C^{14}-labeled amino acids, the cells were lysed and layered onto a 15% to 30% sucrose gradient. The samples were centrifuged at 55,000 × g for 2 hours at 25°C. Fractions were collected from the bottom of the tubes and analyzed for optical density (OD_{260}) and radioactive protein (CPM). (A) Undegraded lysate. (B) Partially degraded lysate.

   This same experiment provided a clue to the nature of the rapidly sedimenting material. The cell extract, which had been "handled less carefully," displayed some "irregularities" in the sedimentation profile (Warner, Knopf, and Rich, 1963, p. 122) (compare Figure 3.14B with Figure 3.14A). Only the partially degraded lysate showed a peak of labeled protein in the region of the 78S ribosome. In another experiment, mild treatment with RNAase shifted all of the labeled protein to the 78S position. These results led to the suggestion that the heavy peak consisted of a multiple ribosome structure held together by RNA, and that this structure be called a polyribosome, or polysome.

   The fragility of polysomes was also demonstrated by repeated pelleting; that is, they were repeatedly centrifuged to the bottom of the tube. Shearing forces were exerted on the pelleted material, both by centrifugal force and during resuspension. The results obtained after three pelleting steps are shown in Figure 3.15. Five peaks of coinciding radioactivity and optical density are evident, with values ranging from 76S to 170S. Fractions from this sucrose gradient were then analyzed by electron microscopy (Figure 3.16). The micrographs provided a convincing demonstration of the existence of discrete-sized polysomes in the various peaks that resulted from the sucrose gradient sedimentation.

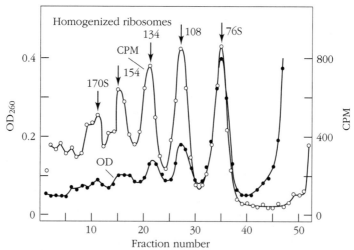

Figure 3.15   Sedimentation rate of protein-labeled polysomes after intentional partial degradation.
Reticulocytes were labeled briefly with H^3-leucine, and a cell lysate was prepared. The polyribosomes were pelleted three times by high-speed centrifugation; each time resuspension was done using a homogenizer. The sample was then layered onto a 5% to 20% sucrose gradient and centrifuged. Estimated S-values of the polysomes are indicated for each peak.

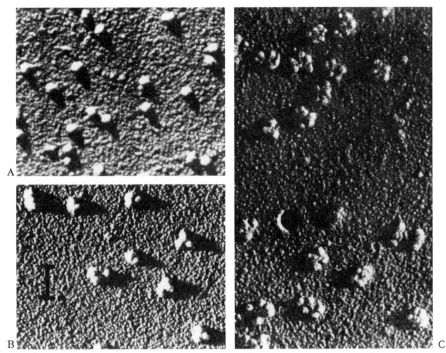

Figure 3.16   Electron micrographs of reticulocyte ribosomes fractionated by sucrose gradient centrifugation. The vertical mark indicates 0.1 μm. (A) Droplet from the 76S peak. (B) Droplet from the 134S peak showing triplets of ribosomes. (C) Droplet from the 170S peak showing several ribosome pentamers.

The 134S peak showed clusters of three ribosomes (Figure 3.16B) and the 170S peak showed clusters of five ribosomes (Figure 3.16C). Later electron microscopic studies more clearly revealed the fine strand of RNA threaded through the 78S ribosomes and led to estimates of the relative sizes of mRNA molecules by simple counting of the number of attached ribosomes.

3.4.3   Evidence for Split Genes and RNA Splicing

Probably the most surprising finding in all the work on mRNA was the discovery of RNA splicing. It was hard enough to come to grips with the wasteful degradation of hnRNA to the smaller sized mRNA. The most probable mechanism that anyone could think of was the removal of segments from one or both ends of the hnRNA, termed *RNA processing*. This idea posed a problem, however, bcause it was quickly learned that the sequences of polynucleotides at both the 5′ and 3′ termini of a specific hnRNA molecule matched those at the termini of the corresponding processed mRNA molecule. How, then, could processing involve the removal of terminal polynucleotide sequences? The first demonstration that RNA processing actually entailed the removal of internal, or intervening, sequences came in 1977 from studies of adenovirus mRNA synthesis. Shortly thereafter, similar findings were observed with cellular hnRNA coding for several distinct proteins.

The elegant procedure for the discovery of cellular hnRNA splicing can be seen in the experiments from the laboratories of Philip Leder and Charles Weissmann that were reported in 1978. These studies concerned the mechanism of processing a 15S precursor to a 10S mRNA molecule that encodes mouse β globin, a subunit of hemoglobin. Pulse-chase analysis had indicated earlier that the 15S RNA was definitely a precursor to the 10S mRNA. It was also known that the 15S RNA, with a length of 1500 nucleotides, corresponded well with the length of the chromosomal gene sequence that coded for β globin. The availability of cloned fragments containing the mouse β-globin genes and purified 15S and 10S RNA made it relatively easy to analyze the mechanism by which the β-globin precursor to mRNA becomes processed. Two techniques that have been described earlier, DNA-RNA hybridization and electron microscopy of nucleic acids, were combined in this work.

Electron microscopic analysis of the hybrid nucleic acid molecule formed when 10S β-globin mRNA is *annealed* to a cloned β-globin gene segment, MβG3, revealed a structure shown in Figure 3.17B. Compare this structure with that of the hybrid formed when the 15S precursor to β-globin mRNA is annealed to MβG3 DNA (Figure 3.17A). With the 15S RNA, a large *continuous* DNA-RNA hybrid was formed, and a single-stranded DNA equal in length to the hybrid was displaced and appeared as a loop. Because single-stranded DNA is less electron dense than a double-stranded DNA-RNA hybrid, they can be readily distinguished in good electron photomicrographs. The loops created when DNA is annealed to RNA are referred to as *R-loops*. Note that the 10S mRNA produced a DNA-RNA hybrid with a very different pattern of R-loops. In this case, a looped intervening segment of DNA was observed between two adjacent R-loops. Many such hybrids were analyzed and the intervening segments were shown to have an average length corresponding to about 550 base pairs (bp).

This powerful R-loop technique permitted these and other researchers to draw an astonishing conclusion. Eukaryotic genes are almost all arranged with such looped or *intervening sequences* (*introns*), which are transcribed but removed from the RNA during the processing of hnRNA into mRNA in the cell nucleus. The structure of the mouse β globin gene is seen in Figure 3.18. It is a *split gene*, having two coding sequences separated by a 550-bp noncoding intervening sequence. When the coding sequences (*exons*) of this gene hybridized with the 10S mRNA derived from it, the noncoding DNA intron had no complementary RNA with which to anneal. The intron thus appeared as a loop

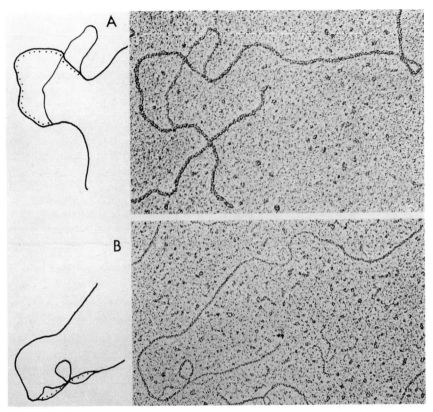

Figure 3.17 Electron micrographs of R-loop structures (140,000✕).
(A) Between 15S RNA and MβG3. (B) Between 10S RNA and MβG3. The
dotted lines in the drawings indicate the positions of the RNA molecules
(140,000✕).

that was located at a site within the DNA-RNA hybrid. The intron loop
caused the DNA-RNA hybrid to appear as two adjacent R-loops rather
than one (see Figure 3.17).

Other genes have been shown to contain more than 10 intervening
sequences, and thus their transcripts require a very large number of cuts
and splices during the formation of their mRNAs. The reason for such
complexity and apparent inefficiency in eukaryotes is currently a hotly
debated subject. Ultimately it may shed considerable light on the evolution of genes and their organization.

3.4.4 The Template for Eukaryotic RNA Synthesis

Yet another surprise was revealed by the scientists who investigated the structure of the template for eukaryotic RNA synthesis. Using
newly devised spreading techniques for viewing preparations of am-

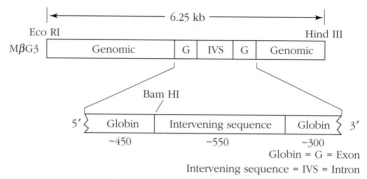

Figure 3.18 Diagrammatic representation of a β-globin gene.
The first line represents the Eco RI-Hind III fragment of MβG3, which was
used to perform the R-loop analysis shown in Figure 3.17. The positions of
the globin mRNA sequences (G) and the major intervening sequence (IVS) are
indicated. The second line represents an enlargement of the sequences that
hybridized to 15S RNA, with the Bam HI site at the codons for amino acids
98–100 indicated. Below the line are the approximate number of base pairs
in each G and IVS.

phibian oocyte nucleoli in the electron microscope, Oscar Miller and
Barbara Beatty in 1969 obtained some beautiful micrographs catching
ribosomal RNA (rRNA) in the act of synthesis (Figure 3.19). The photos
showed that the template for rRNA consists of circular deoxyribonucleo-
protein (DNP) molecules that contain highly active, multiple copies of
rRNA genes, each of which is separated from its neighboring genes by
inactive spacer segments of DNP. The feather pattern created by the
nascent RNA molecules projecting outward from the DNP template
indicates the direction of transcription, with the shortest RNAs being
nearest the site of initiation for each transcription unit, or rRNA gene.

Shortly after this work was published, the structure of *chromatin*
became clarified (see Chapter 1). The basic proteins, *histones*, were
shown to be organized into particles called *nucleosomes* around which
the DNA was wrapped, giving the appearance in electron micrographs of
beads on a string. For many years it was dogmatically believed (with the
usual iconoclastic exceptions) that histones act as repressors (inactiva-
tors) of gene function, preventing certain genes from being transcribed
by RNA polymerase into mRNA. According to this belief, nucleosomes
should be present only in the regions of inactive genes, and not in those
genes that are actively engaged in mRNA synthesis.

A thorough electron microscopic analysis of chromatin obtained
from *Oncopeltus fasciatus* (milkweed bug) convincingly destroyed this
dogma. Some micrographs obtained by Victoria Foe, Linda Wilkinson,
and Charles Laird, published in 1976, are shown in Figure 3.20. The
chromatin appearing between nascent mRNA transcripts was shown to
contain the beaded morphology of typical nucleosomes, with diameters

Figure 3.19   Extrachromosomal rRNA genes isolated from an oocyte of *N. viridescens*.
The featherlike pattern of RNA fibrils extends from a circular DNA molecule, and it clearly demonstrates the great efficiency of rRNA synthesis on this template.

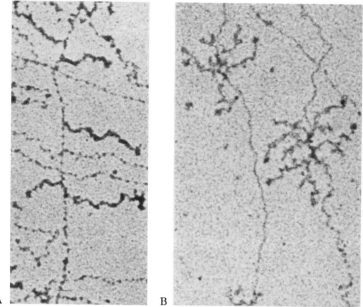

Figure 3.20 Comparative morphologies of chromatin from *O. fasciatus* (67,000✕).
(A) Growing chains of mRNA extend from "beaded" (nucleosome-containing) chromatin. (B) Growing chains of rRNA extend from unbeaded DNA.

of 135Å ± 24Å (Figure 3.20A). The frequency of nucleosomes between the extending RNA fibers was about 0.7 times that of fiber-free chromatin. Nevertheless, this was still a sufficient number of nucleosomes to render invalid the argument that histones inactivate gene expression.

Examination of ribosomal transcription units having relatively few RNA fibers permitted visualization of the DNA template for rRNA (Figure 3.20B). This template was shown to be unbeaded, with a mean diameter of 73 ± 17 Å. Nevertheless, the thickness of the DNA in relation to that of free DNA (20 Å) and its staining properties with phosphotungstic acid suggested that it was associated with proteins. Whether or not the proteins associated with the rRNA genes were histones could not be determined from these studies.

One picture can indeed be worth a thousand words when it can be readily interpreted. This elegant study leaves us with little doubt: the template for mRNA synthesis in eukaryotes is chromatin, not naked DNA. But as always, many questions are raised for every one that is answered. One big mystery posed by these discoveries is the biochemical difference between active and inactive chromatin; or, phrased another way, what turns genes on and off?

Discussion Questions

1. How was it shown that heavy and light ribosomes do not aggregate or exchange RNA components with each other?

2. How was it shown that most of the RNA synthesized after T2 infection of *E. coli* was associated with 70S ribosomes?

3. How was it demonstrated that some of the RNA synthesized after T2 infection has sequences that are homologous (complementary) to T2 DNA, but not to *E. coli* DNA?

4. How can you account for the observation that the T2 DNA-RNA hybrid had the same density as denatured T2 ^{3}H-DNA?

5. Could a CsCl density gradient be used to determine the sedimentation rate of RNA? Explain your answer.

6. In a pulse-chase experiment, 60 percent of the pulse-labeled RNA became acid soluble during the first five minutes after the ^{3}H-uracil was removed, but 40 percent of the total ^{3}H-RNA remained acid precipitable for more than an hour during the chase. What percentage of the pulse-labeled RNA appears to be mRNA and what is its *maximum* half-life by this assay procedure (acid precipitation)?

7. How did the R-loop technique lead to the discovery of introns (intervening sequences) and exons?

References

Astrachan, L., and E. Volkin (1959). Effects of chloramphenicol on ribonucleic acid metabolism in T2-infected *Escherichia coli*. *Biochimica et Biophysica Acta, 32:*449.

Black, D. L., B. Chabot, and J. A. Steitz (1985). U2 as well as U1 small nuclear ribonuclearproteins are involved in premessenger RNA splicing. *Cell, 42:*737.

Brenner, S., F. Jacob, and M. Meselson (1961). An unstable intermediate carrying information from genes to ribosomes for protein synthesis. *Nature, 190:*576.

Cairns, J., G. S. Stent, and J. D. Watson (eds.) (1966). *Phage and the Origins of Molecular Biology.* Cold Spring Harbor Laboratory of Quantitative Biology.

Cantor, C. R., and P. R. Schimmel (1980). *Biophysical Chemistry, Part II: Techniques for the Study of Biological Structure and Function.* New York: W. H. Freeman.

Doty, P., J. Marmur, J. Eigner, and C. Schildkraut (1960). Strand separation and specific recombination in deoxyribonucleic acids: physical chemical studies. *Proceedings of the National Academy of Sciences USA, 46:*461.

Fawcett, D. W. (1981). *The Cell* (2nd ed.), pp. 72–73. Philadelphia: W. B. Saunders.

Foe, V. E., L. E. Wilkinson, and C. D. Laird (1976). Comparative organization of active transcription units in *Oncopheltus fasciatus. Cell, 9*:131.

Hall, B. D., and S. Spiegelman (1961). Sequence complementarity of T2-DNA and T2-specific RNA. *Proceedings of the National Academy of Sciences USA, 47*:137.

Higgs, D. R., S. E. Y. Goodbourn, J. Lamb, J. B. Clegg, D. J. Weatherall, and N. J. Proudfoot (1983). α-thalassaemia caused by a polyadenylation signal mutation. *Nature, 306*:398.

Jacob, F., and J. Monod (1961). Genetic regulatory mechanisms in the synthesis of proteins. *Journal of Molecular Biology, 3*:318.

Judson, H. F. (1979). *The Eighth Day of Creation.* New York: Simon & Schuster.

Maniatis, T., E. F. Fritsch, and J. Sambrook (1982). Propagation and maintenance of bacterial strains and viruses. In *Molecular Cloning.* Cold Spring Harbor Laboratory of Quantitative Biology.

Marmur, J., and D. Lane (1960). Strand separation and specific recombination in deoxyribonucleic acids: biological studies. *Proceedings of the National Academy of Sciences USA, 46*:453.

Meselson, M., and F. W. Stahl (1958). The replication of DNA in *Escherichia coli. Proceedings of the National Academy of Sciences USA, 44*:671.

Miller, O. L., Jr. (1981). The nucleolus, chromosomes, and visualization of genetic activity. *Journal of Cell Biology, 91*:15S.

Miller, O. L., Jr., and B. R. Beatty (1969). Visualization of nucleolar genes. *Science, 164*:955.

Pardue, M. L., and J. G. Gall (1969). Molecular hybridization of radioactive DNA to the DNA of cytological preparations. *Proceedings of the National Academy of Sciences USA, 64*:600.

Perry, R. P., E. Bard, B. D. Hames, D. E. Kelley, and U. Schibler (1976). The relationship between hnRNA and mRNA. *Progress in Nucleic Acid Research in Molecular Biology, 19*:275.

Puckett, L., and J. E. Darnell, Jr. (1977). Essential factors in the kinetic analysis of RNA synthesis in HeLa cells. *Journal of Cell Physiology, 90*:521.

Ruskin, B., A. R. Krainer, T. Maniatis, and M. R. Green (1984). Excision of an intact intron as a novel lariat structure during pre-mRNA splicing *in vitro. Cell, 30*:317.

Sagik, B. P., M. H. Green, M. Hayashi, and S. Spiegelman (1962). Size distribution of "informational" RNA. *Biophysical Journal, 2*:409.

Tilghman, S. M., P. J. Curtis, D. C. Tiemeier, P. Leder, and C. Weissmann (1978). The intervening sequence of a mouse β-globin gene is transcribed within the 15S β-globin mRNA precursor. *Proceedings of the National Academy of Sciences USA, 75*:1309.

Warner, J. R., P. M. Knopf, and A. Rich (1963). A multiple ribosomal structure in protein synthesis. *Proceedings of the National Academy of Sciences, USA 49*:122.

4

Cell Differentiation, Aging, and Cancer: Cell Culture Studies

Often a technique becomes so well established and useful for such a wide variety of purposes that its significance is overlooked or taken for granted. This is the case with the method referred to as *tissue culture* or, more appropriately, as *cell culture*. Students are no longer surprised to learn that animal and plant cells can be grown in culture, in many cases indefinitely, and still retain many of their differentiated properties. Yet even today the ability of researchers to maintain healthy cells in culture, free of bacteria, yeast, molds, and other common contaminants, is highly valued. Laboratory technicians with this skill are in great demand around the world.

It is impossible to grasp the art and technology of cell culture without spending considerable time in the laboratory, but we can understand some of the most important biological questions that are being investigated by means of such technology without setting foot in a lab. One of the most intriguing problems in the field of developmental biology is how an egg manages to become a chicken, elephant, or any other particular animal. This process—the process of cell differentiation—can be studied by means of animal cells grown in culture.

We'll ignore the egg cell (oocyte) for the present and consider instead some other cells commonly grown in the laboratory: neural (nerve) cells, myoblasts (muscle), and lymphocytes (blood), all of which exhibit the fascinating capability of differentiating in cell culture.

A heterogeneous collection of cells may be obtained from ganglia, enriched for specific cell types such as neurons and glia, then used to study the effects of specific factors on nerve cell functions. The first transplant of cells from the muscle into ·a growth medium—the so-called *primary culture*—can be used to create an established, or permanent, line of proliferating myoblasts that retain the ability to fuse with neighboring cells of the same type. Such fused cells then go on to produce muscle-specific proteins. Chemicals can stimulate cells in the lymphocyte system to leave the resting stage of the cell cycle and enter a proliferative stage that involves DNA replication and mitosis. Such properties of lymphocytes make them the focus of investigations into the basis of cancer.

Cells in culture provide a wonderful system to study not only cancer but also the aging process. Cancer and aging are more closely related than they may seem. Animal cells grown in culture actually age in the sense that after a certain number of cell divisions, all the cells lose their ability to divide in a fairly synchronized manner. The loss of replicative capacity is a function of the species of the specimen from which the cells were obtained as well as of its age. This is referred to as the *Hayflick phenomenon*, after its discoverer, Leonard Hayflick. His findings have interesting implications for the process of aging, as we shall see.

As one might predict, many types of cancer cells are immortal in culture. That is, they can divide indefinitely without exhibiting the Hayflick phenomenon. In efforts to study the mechanism by which normal cells become converted to cancerous (tumorigenic) cells, several common and useful assays have been developed. They rely on observed changes (transformations) in cellular growth and morphology. How does one decide whether *transformed cells* are indeed tumorigenic? Here we go a bit beyond cell culture with the use of "nude" mice to determine whether the immortal cells can grow into a tumor. The nude mouse is a developmental mutant that has lost its ability to produce a thymus gland. For this reason the poor animal has a very defective immune defense system. It may thus serve as a host for the growth of tumorigenic cells originating in any species, including our own.

A very important goal in the field of cancer research is the identification of genes that convert a normal cell into a tumorigenic cell (oncogenes) and genes that suppress the phenotypic properties of a tumorigenic cell (tumor suppressor genes). Such studies require the technical ability to identify and carefully examine each of the chromosomes in a cell. This technique, referred to as *karyotypic analysis*, has been used in some very recent classic experiments concerning the basis of cell immortalization and cancer. We'll investigate methods for the isolation of specific genes in Chapter 5.

4.1 Cellular Differentiation in Vitro

4.1.1 Nerve Cells (Neurons and Glia)

One of the first great experiments to use tissue culture methodology was performed with nerve tissue in 1907. Ross Harrison cultured pieces of tissue from the medullary tube region of frog embryos in clots of frog lymph. Not only did the cells survive for several weeks, but nerve fibers grew out from some of the cells. Of course tissue culture technology has become far more sophisticated today. Methods for obtaining homogeneous cell populations capable of growing and differentiating in a medium that is completely defined in terms of its chemical composition have been achieved for many cell types. A fine example of such progress can be seen in some experiments conducted by Silvio Varon and his colleagues.

Starting with neonatal (1 to 48 hours old) mouse dorsal root ganglia, located in the spinal cord, these investigators obtained cultures of either pure Schwann cells or Schwann cells associated with neurons. Schwann cells are specialized glial cells present in peripheral nerves that produce a myelin sheath. The sheath wraps around axons, insulating them and helping them to propagate nerve impulses more efficiently. Glial cells (glia) provide many essential metabolic and structural functions for the nervous system, both during its process of development and throughout life. It is very important to be able to culture the cells found in the nervous system (the neurons and many types of glia), and to examine their properties both in purified homogeneous cell culture and in mixed populations where the different types of cells have the opportunity to interact with one another as they do in vivo.

The dissected ganglia are first dissociated into their component cells so that the cell types can be separated. Mild digestion with enzymes such as trypsin hydrolyzes proteins on the cell surface and helps to dissociate the tissue into a suspension of small cell aggregates and single cells. The experimenter further disperses the cells by straining them through metal or nylon sieves or, more commonly, by repeatedly aspirating them in and out of a small-bore pipet. From a single neonatal mouse dorsal root ganglion about 7000 cells can be obtained, of which 45 percent are neurons and most of the remainder are Schwann cells. The cells are diluted into growth medium and grown in small plastic dishes (35-mm diameter) coated with an adhesive protein, collagen, that helps the cells attach to the surface of the dish. The cells are grown in incubators kept at 37°C in a humidified 5 percent CO_2–95 percent air atmosphere, and the growth medium is replaced every two days. Of course, absolutely sterile conditions must be maintained throughout this entire procedure to avoid contaminating the cultures.

A variety of tricks can be used to enrich tissues for specific types of cells that can be induced to perform feats that display their differentiated properties. In this project, Varon and his colleagues seeded the disso-ciated ganglia onto plastic dishes that were not coated with collagen. After 2.5 hours, only nonneurons were attached firmly to the dishes. The unattached cells—about 85 percent of the total—were thus quickly enriched in neurons. This cell fraction was then placed again into growth medium in dishes coated with collagen. The growth medium contained purified nerve-growth factor (NGF) and fetal calf serum (FCS). Neurite formation, which is the sending out of axons and den-drites from the neuronal cells, is known to depend on the presence of NGF. FCS contains the factors that support the proliferation of nonneur-onal cells. Recall that neurons are terminally differentiated cells that have lost their ability to divide. Photographs of these cultures taken after various periods of growth are shown in Figure 4.1.

By 24 hours (Figure 4.1A), most of the neurons extended one or more neurites, some displaying considerable branching. A few "small-flat" and spindle-shaped Schwann cells were present. The cultures were then transferred to a medium containing NGF, but no FCS. By nine days (Figure 14.1B), a very profuse neuritic network was extended. Schwann cells greatly increased in number, and virtually the entire neuritic net-work was occupied by small-flat and spindle-shaped Schwann cells. At this time some cultures were transferred to a medium containing FCS in place of NGF for another 24 hours. The morphology of these cultures was greatly altered, as we can see when we compare Figure 4.1B (before FCS was added) and Figure 4.1C (the same microscopic field 24 hours after FCS was added). The addition of FCS caused the Schwann cells to

---

Figure 4.1 Cell morphologies within cultures enriched for neuron-satellite cells (NS).
(A) Twenty-four hours in vitro (immediately before shifting to medium plus NGF). Note the profuse network of relatively short neurites from some neurons (single arrow), and some Schwann cells beginning to form apparent associations with neurites (double arrows). (B) Nine days in vitro (medium plus NGF). Note that the entire neuritic region appears occupied by tightly associated spindle (single arrow) and small-flat (double arrow) Schwann cells. Large-flat fibroblasts, which maintain only a random association with neurites, are rarely seen (not shown). (C) Same field as in (B) but 24 hours after replacement of medium plus NGF with medium plus fetal calf serum (FCS). Note that neuronal survival has continued, and Schwann cells appear to have dissociated from the neurites, leaving a naked neuritic network above the Schwann cell layer (arrow). The Schwann cells have assumed a much more flattened morphology, and cell numbers appear to have increased above the level of the day before.

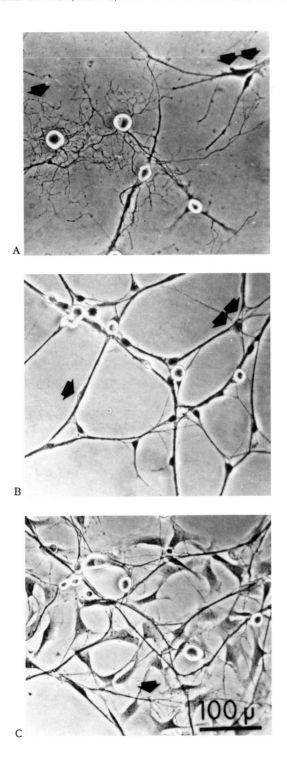

become flatter and lose their close association with the neurites, which appear as "a naked network overlying the flat cell bed" (Manthorpe, Skaper, and Varon, 1980, p. 467).

It is apparent that fetal calf serum not only contains mitogens that support the proliferation of mouse Schwann cells, but also contains factors that cause a rapid change in the morphology of Schwann cells and cause them to lose their association with neurites. The researchers predicted that these findings "may have considerable relevance to the consequences of nerve lesions *in vivo*, where injury to local blood vessels might similarly lead to an interruption of glial/axon communications" (Manthorpe, Skaper, and Varon, 1980, p. 467).

This work has led to a very exciting area of current research, the identification and biochemical characterization of factors that are required for the survival of neurons, as distinct from their production of neurites. The factors required for neuron survival are called *neuronotrophic factors* (*NTF*), while the ones required for outgrowth of neuronal processes (axons and dendrites) are called *NPF*, for *neurite-promoting factors*.

The results of an experiment by Varon and colleagues (1988) illustrating the effects of NTF and NPF on cultured neurons from chick embryo ciliary ganglia are shown in Figure 4.2. The neurons failed to survive the absence of both NTF and NPF (0/0), survived but did not produce neurites when only NTF was present (NTF/0), began to grow neurites but later died when NPF but not NTF was in the medium (0/NPF), and survived and produced profuse neurites in the presence of both factors (NTF/NPF).

The effect of NGF, NTF, NPF, and other yet-to-be discovered neural factors on processes such as learning and memory and on diseases that affect the functioning of the brain and of the neuromuscular system can only be speculated on at this time. The field of neurobiology obviously has great potential and is attracting outstanding scientists.

4.1.2 Muscle Cells (Myoblasts)

The visual effect of cell differentiation, or increasing specialization, in culture is even more dramatic with muscle cells, or *myoblasts*. David Yaffe and his coworkers at the Weizmann Institute in Israel were the first to establish permanent lines of rat myoblasts. The cells can divide indefinitely in culture when their density is kept sufficiently low. When their density on the plastic dishes is high, the cells carry out their predetermined functions as differentiated muscle; that is, they fuse with one another and form cross-striated multinucleated muscle fibers (*myotubes*).

The experimenters prepared primary skeletal muscle cultures from the thigh muscle of newborn rats. As with all primary cultures derived

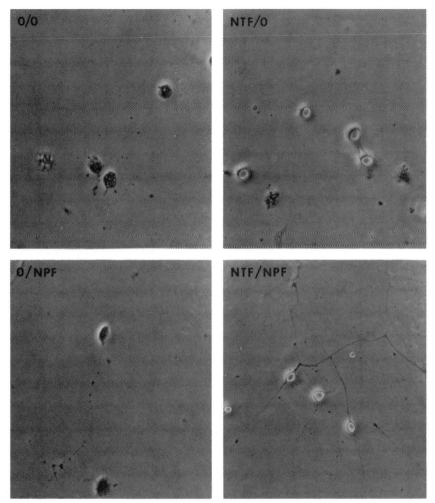

Figure 4.2 Separate effects of NTFs and NPFs on cultured neurons. Purified neurons from chick ciliary ganglia were cultured on plates in a medium with and without NTF and/or NPF (indicated at upper left of each photo). The neurons survive only in the presence of NTF, and they form neurites only in the presence of NPF.

from an organ or tissue, they obtained a very heterogeneous mixture of cell types. They grew the cells in a rich medium supplemented with 10 percent horse serum and 1 to 3 percent chick embryo extract. Thus the medium was complex and of unknown chemical composition.

To select myoblasts from the primary cultures, Yaffe and his associates removed the cells from the dishes by mild enzymatic treatment with the protease trypsin shortly before the cells fused. They then placed the cells into untreated plastic petri dishes; that is, dishes not coated with collagen. After a short time (40 to 60 minutes), most of the fibro-blastic and epithelial cells (so called because of their spindle-shaped and cobblestone morphologies, respectively) became attached to the plastic surface. Most of the myoblasts remained floating in the medium. The experimenters removed the myoblasts and placed them in colla-gen-treated petri dishes. They repeated this selective passaging each time the cells approached high density and were about to fuse, thereby obtaining a population greatly enriched in myoblasts.

To obtain a genetically homogeneous cell population, it is always necessary to isolate a *clone*; that is, a population that arises from a single cell. Yaffe and his colleagues got their clone by placing a sterile glass ring over a small colony of cells and sealing it to the surface of the plate with silicon grease. They detached the colony of cells with a drop of trypsin, then aspirated off the cells and diluted them in growth medium for further culture. Cells derived from a clone and maintained indefi-nitely in the laboratory are referred to as a *cell line*. Figure 4.3 shows colonies derived from single cells of a myoblast line called L6.

Obtaining lines of myoblastic cells was no simple matter. These investigators initiated more than 20 experiments in an effort to establish such cell lines. Cell multiplication usually ceased after three to five passages, and the line was lost. The generation time increased markedly as the cells were passaged, so that five passages required the care and feeding of cells for about eight weeks. Fortunately, in six experiments the cultures recovered at later passages, and growth rates became similar to those of primary cultures. The experimenters suggest that this "recov-ery" took place in one or a very few cells on a dish. It is plausible that these rare cells that gave rise to myogenic lines were the result of specific mutations that occurred in the animal. However, other interpre-tations seem equally likely.

Later the Yaffe group described some of the biochemical changes that took place in their muscle cell lines after they differentiated in culture. As with muscle tissue in vivo, the activities of several enzymes change markedly after myoblasts (undifferentiated mononucleated cells) have differentiated into myotubes (muscle fibers), as Figure 4.4 clearly shows. Here the primary muscle culture is from rat thigh and the enzyme is glycogen phosphorylase (GPh). Using a cytochemical assay for GPh, in which enzyme activity results in a bluish-gray staining, the investigators found that the enzyme was localized predominantly in the larger myotubes. Little or no stain appeared in the myoblasts.

They then assayed three muscle-specific enzymes as a function of culture time with the L6 myoblast line (Figure 4.5). The activity of each enzyme — GPh, creatine phosphokinase (CPK), and myokinase (MK) — began to increase after cell fusion started. Because cell fusion was not synchronized, this process and the increases in enzyme activities stretched over a period of more than 10 days. As a control, the experi-menters prevented cell fusion by lowering the Ca^{++} ion concentration in

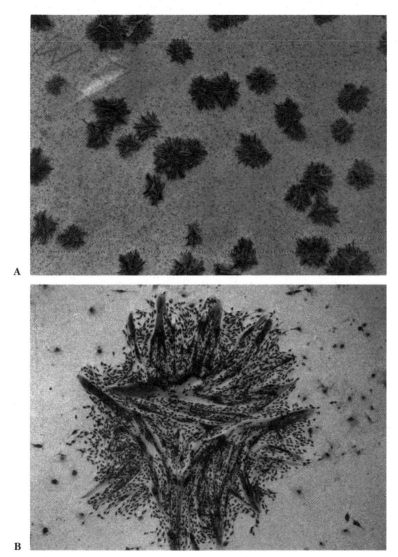

Figure 4.3 Muscle-forming colonies derived from single cells of line L6. Cells were fixed 19 days after plating. (A) Magnification 4.5X. (B) Higher magnification (31.5X) of a single colony in (A), showing the internal organization of the colony. The mononucleated cell types appear quite homogenous in morphology.

the medium. They found no increase in these enzymatic activities. To show that the reduction in Ca^{++} had not simply killed the cells, they added Ca^{++} back to the cultures. The cells rapidly fused and the three enzyme activities increased as before.

Permanent cell lines such as the myoblasts offer many advantages over animals for the study of differentiation. Questions concerning the

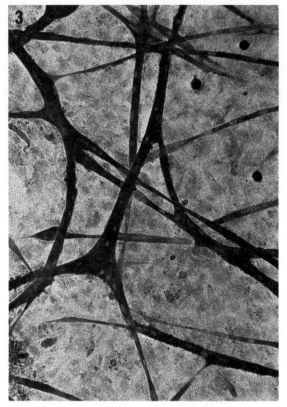

Figure 4.4 Cytochemical demonstration of glycogen phosphorylase activity in differentiating primary muscle cell cultures (140X).

molecular basis of the differentiation process, including the accompanying changes in gene expression that lead to elevations in the activities of specific enzymes, can be more readily addressed with homogeneous cultured cells. Established cell lines also offer the opportunity to isolate mutants that affect the process of differentiation in specific ways. Thus the ultimate goal of understanding the molecular basis of differentiation becomes much more readily attainable.

4.1.3   Lymphocytes (T Cells and B Cells)

Unlike the terminally differentiated neurons and muscle fibers described earlier, certain types of highly differentiated lymphocytes such as T cells and B cells possess the innate ability to proliferate in response to chemical signals called *mitogens*. This is a fundamental characteristic of their differentiated state. Activation of T cells and B

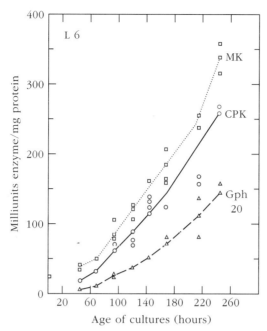

Figure 4.5 Enzyme activities in cultures of the myogenic cell line L6 as a function of age.
MK = myokinase; CPK = creatine phosphokinase; G.Ph. = glycogen phosphorylase.

cells results from exposure to a wide variety of mitogens, which can be obtained from plants, bacteria, and other sources. It is now realized that the response of these lymphocytes to mitogens is related to a major differentiation property—the ability to proliferate rapidly upon interaction with a specific *antigen* (a substance foreign to the body). This key step in the body's immune response results in the production of a clone of lymphocytes in which every cell produces antibodies directed against the antigen.

In 1962 Michael Bender and David Prescott reported what has become a standard procedure (with modifications, of course) for the isolation and culture of human peripheral blood leukocytes, including lymphocytes. It is called the buffy coat separation method.

> Sterile blood (5 ml/culture) is drawn by venipuncture and placed in sterile, screw cap 15-ml centrifuge tubes containing 0.1 ml of commercial heparin (1000 units/ml). The tubes are spun at maximum speed in an International clinical centrifuge for 10 min. at room temperature. The leukocytes form a dense layer (buffy coat) lying on top of the packed erythrocytes. The serum, buffy coat, and the top 1 mm of erythrocytes are withdrawn with a sterile Pasteur pipet and added to a milk dilution bottle containing the culture medium. The

cell clumps are then broken up by vigorous pipeting. (Bender and Prescott, 1962, p. 221)

The unsealed cultures are grown in incubators at 37°C in a 5 percent CO_2–95 percent air atmosphere. The growth medium contains antibiotics (penicillin and streptomycin) to help prevent contamination. Unlike the cells described earlier, lymphocytes grow in suspension, not attached to a surface. The appearance of lymphocytes and a macrophage obtained from the peritoneal cavity of a mouse is seen in the scanning electron micrograph shown in Figure 4.6.

One of Bender and Prescott's experiments was designed to deter-

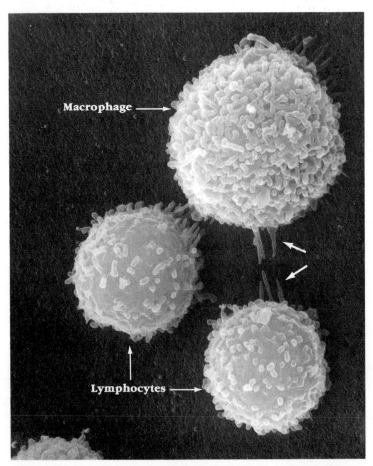

Figure 4.6 Scanning electron micrograph of a peritoneal macrophage and two lymphocytes.
The surface of the macrophage is entirely covered by short folds and microvilli. In contrast, the lymphocytes show expanses of smooth surface between sparse microvilli. These features generally serve to distinguish these two cell types.

mine whether an individual lymphocyte can undergo more than one division cycle in cell culture. *S-phase* lymphocytes (those that synthesize DNA) were labeled with ^{3}H-thymidine for 30 minutes, starting 48 hours after exposure to the mitogen phytohemagglutinin (PHA). The cells were then transferred to a nonradiolabeled medium. After tagging the newly synthesized DNA by this pulse-labeling procedure, Bender and Prescott treated separate cultures with colchicine from 67 to 151 hours after they had added PHA. They always added colchicine, a drug that arrests cells in the metaphase stage of mitosis, five hours before they "fixed" a sample of cells. (The fixing of cells will become clear in a moment.) They fixed a separate sample of cells every 12 hours from 72 to 156 hours after the addition of PHA. A diagram illustrating the complex protocol of this experiment is shown in Figure 4.7.

To fix the cells for microscopy, Bender and Prescott centrifuged the cells and washed them with a hypotonic solution, then centrifuged them again. After they had removed the hypotonic solution, they added a fixative to the cells in the pellet. The fixative contained methyl alcohol and glacial acetic acid in a ratio of 3:1 by volume. After one-half hour they suspended the cells in the fixative again, washed them several times by centrifugation, and then mounted them by placing several drops on the clean wet surface of a slide. After air drying, the slides were ready for autoradiography and visualization of clearly separated metaphase chromosomes (Figure 4.8). Now let's examine the results of the experiment.

First mitotic divisions could be easily distinguished from all subsequent divisions as follows. In the first metaphase after the DNA was pulse-labeled, both chromatids of all chromosomes were labeled. In all subsequent mitoses, however, not more than one chromatid of a given

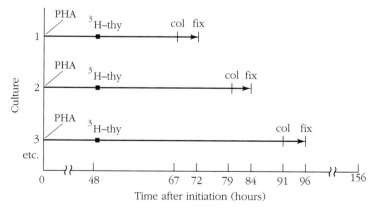

PHA: phytohemagglutinin
thy: thymidine
col: colchicine

Figure 4.7 Experimental protocol of Bender and Prescott.

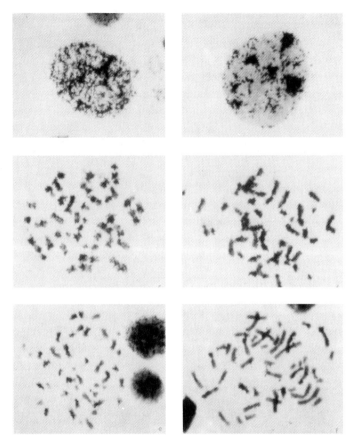

Figure 4.8 Autoradiographs of tritiated thymidine-labeled human peripheral leukocytes from cultures fixed 72 to 156 hours after mitotic stimulation. (A) Large and small monocytes showing a moderate level of labeling. (B) Localized labeling in a large monocyte. (C) Chromosomes of a first post-DNA-labeling metaphase. Both chromatids of each chromosome are labeled. (D) Chromosomes of a second post-DNA-labeling metaphase. Only one chromatid of each chromosome is labeled. (E) Chromosomes of a third post-DNA-labeling metaphase. Only about half of the chromosomes contain a labeled chromatid. (F) Chromosomes of a fourth post-DNA-labeling metaphase. Only about one-fourth of the chromosomes contain a labeled chromatid.

pair would remain labeled. Recall that ^{3}H-thymidine was present for only 30 minutes in the entire experiment, beginning at 48 hours after the addition of the mitogen PHA, and that each chromatid contains double-stranded DNA, which replicated semiconservatively during the S phase before mitosis. Thus, in the second metaphase after DNA labeling, one chromatid of a pair would remain labeled, but not the other. In the third metaphase after the DNA was labeled, only about half of the duplicated chromosomes (one-fourth of the total chromatids) would still be labeled, and so on for subsequent mitoses. If this seems confusing, a review of mitosis and the cell cycle would be helpful.

Labeled first, second, third, and fourth mitotic divisions can be seen in the autoradiographs taken of the metaphase-arrested lymphocytes (Figure 4.8). Therefore Bender and Prescott conclude that an individual lymphocyte can divide in culture at least four times.

Experiments of this nature helped to substantiate the current view that certain lymphocytes in the blood await a mitogenic signal provided by some foreign antigen, and then they proliferate rapidly in response to this stimulus. In this way a clone of lymphocytes capable of recognizing one specific antigen is generated in an animal. Each cell in the clone produces antibodies that are directed against the foreign antigen that acted as the mitogen, thereby enabling the host animal to detect and destroy foreign invaders such as bacteria and viruses.

The creation of immortal lymphocytes was another very important application of cell culture technology that facilitated rapid progress in the field of immunology. By fusing lymphocytes with tumorigenic cells, one can obtain clones that produce a single highly specific antibody. Such clones give rise to so-called *hybridoma* cell lines, and the antibodies they produce are called *monoclonals*. The use of these *monoclonal antibodies* offers profound possibilities for therapy and diagnostics, and research is proceeding rapidly in this area.

4.2 Cellular Aging, Immortalization, and Cancer

4.2.1 Cellular "Aging" in Culture

One of the most remarkable characteristics of animal cells grown in culture is that most of them lose their ability to divide and eventually die. What initially appeared to be an artifact of inadequate cell culture technology was first shown to be a property of the cells themselves by Leonard Hayflick, and the phenomenon has given rise to an area of cell biology called cytogerentology. Although none of the many hypotheses offered has satisfactorily explained the so-called Hayflick phenomenon, its inherent interest as well as its relevance to the topics of cell differentiation and cancer make cellular aging worthy of consideration.

Figure 4.9 shows the results of one of Hayflick's major experiments. The data shown here, obtained with human fetal lung cells (strain WI-44), were also obtained with other human cells derived from fetal, neonatal, and adult tissues. After cells become established in culture (phase I, not shown), they proliferate rapidly at a relatively constant rate (phase II). Figure 4.9 shows that the cell count remains constant and high after each serial passage at a 2 : 1 ratio. In other words, after a certain fixed period—say, three days—one dish of cells was split into

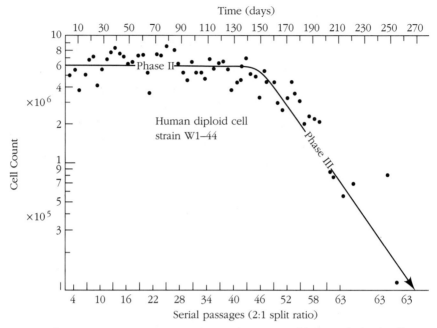

Figure 4.9 Cell counts determined at each passage of human diploid cell strain WI-44.
The plateau (phase II) indicates a constant doubling time for the cell population. During phase III, the doubling time of the population increases exponentially.

two dishes, and the cells were then permitted to continue growing. At a fairly precise serial passage number that is characteristic of each cell type (about 43 in this experiment), the cells enter a phase of exponentially declining proliferative capacity (phase III) until eventually the entire culture is lost. It should be noted that recent experiments indicate that under appropriate conditions, phase III need not end in the death of the cells even though they stop dividing.

In critical control experiments, Hayflick demonstrated that the loss of proliferative capacity (*senescence*) of the old cells was in no way related to the conditions of the culture. Old and young cells, identifiable by karyotypic (chromosomal) markers such as *Barr bodies* (highly condensed X chromosomes), were mixed and grown together on the same dish. The young cells continued to grow while the old cells reached their appointed number of cell divisions and stopped dividing.

The average number of population doublings achieved before senescence was about 50 for cells derived from human fetal lung, but only about 20 for cells obtained from human adult lung. In a thorough study

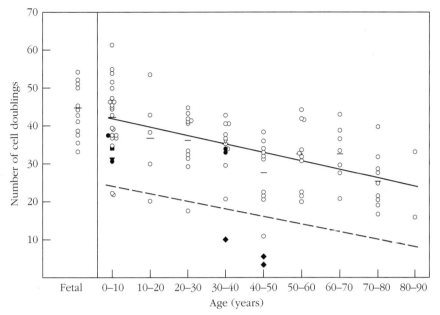

Figure 4.10   The cumulative number of doublings achieved by human skin fibroblast cells in culture as a function of the age of the donor.
The calculated linear regression line (solid line) is drawn between the first and ninth decades and has a regression coefficient of -0.20 ± 0.05 standard deviation cell doublings per year with correlation coefficient of -0.50. The dashed line is the lower 95% confidence limit for the regression line.

of this observation, George Martin and his colleagues measured the cumulative number of doublings achieved by human skin fibroblasts in culture as a function of the donor's age. They found a linear decrease in the life span of the cells in culture as the donor became older (Figure 4.10). This is the most convincing type of evidence that there is actually some relationship between the aging of cells in culture and the aging of the human body.

This point has been much debated, but the controversy should not cause us to overlook another important observation. Studies of this sort have also been conducted with cells obtained from individuals with progeria (Hutchinson-Gilford syndrome) and Werner's syndrome, diseases resulting in accelerated aging. Children with progeria often express the symptoms of old age by the time they reach their ninth year, exhibiting a slowing of growth, graying and loss of hair, atherosclerosis, and osteoporosis. Similar symptoms appear in people with Werner's syndrome, but generally not until they reach their 30s or 40s. Fibroblasts from subjects with these disorders had a considerably shorter life span than normal in culture. Only 2 to 10 doublings occurred, rather than the

20 to 40 with cells derived from normal people of the same age. These findings strongly support the view that studies of the aging of cells in culture may offer valuable insights into the mechanisms of the aging process taking place in our bodies.

4.2.2   Immortalization of Cells in Culture

Having just seen findings demonstrating that animal cells have a finite life span in culture, with a limited, predetermined number of cell divisions, you may find it paradoxical to take up the immortalization of cells. But perhaps it will come as no surprise to learn that most (but not all) cells become immortal when they are made tumorigenic, by either virus infection, radiation, or chemical carcinogens. Tumorigenic cells, however, are not the only ones that can attain immortality. Recall the case of the rat myogenic line L6, described earlier (Figure 4.3). These normal cells can be passaged indefinitely and still retain their capacity to differentiate into myotubes.

Before examining how immortal cells can be tested for tumorigenicity, let's first consider some standard classic methods for isolating *transformed cells*—cells that have lost at least one type of growth control in culture. Both DNA and RNA viruses are capable of transforming cells of virtually all types. Of the many types of RNA viruses, the major ones that transform cells are members of the retrovirus group. These viruses carry within their coat an enzyme, *reverse transcriptase*, that copies the viral RNA genome into a DNA copy. This DNA then integrates into the cell's chromosomal DNA, causing a permanent change in the cell with respect to both its genotype and its phenotype. A similar, albeit more direct, process occurs with the DNA viruses. All four of the main groups of DNA viruses—the poxviruses, the adenoviruses, the papova viruses, and the herpes viruses—contain oncogenic members. Among these viruses, only the poxviruses are not known to give rise to permanent cell lines in culture.

Howard Temin and Harry Rubin used a process called *focus formation* to quantitate the *titer* (infectious virus concentration) of the retrovirus Rous sarcoma virus (RSV). In the early 1900s, Peyton Rous identified this virus as the causative agent in the appearance of fibroblastic solid tumors (*sarcomas*) in chickens. Temin and Rubin assayed RSV by infecting monolayers of chick embryo secondary cultures (passaged once from the primary cultures). After waiting briefly for the viruses to attach to the cells, they added growth medium containing 0.6 percent agar over the infected cells. After five to seven days they added a dye called neutral red to make the foci easier to count. Each focus is a clone of cells that originates from the transformation of one cell by a single virus. After

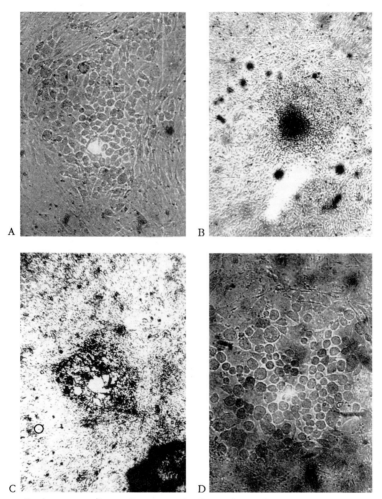

Figure 4.11 Assay of transformation by focus formation.
(A) Focus of Rous sarcoma transformed cells formed on a secondary chick embryo culture; unstained (80X). (B) Focus of Rous sarcoma transformed cells formed on a secondary chick embryo culture; lightly stained with neutral red (20X). (C) Focus of Rous sarcoma transformed cells formed on a secondary chick embryo culture; heavily stained with neutral red (20X). (D) Focus of Rous sarcoma transformed cells formed on a clonal population of chick embryo fibroblasts; unstained (80X).

removing the dye-containing solution, Temin and Rubin scanned the plate for foci through an inverted microscope.

As Figure 4.11 indicates, the foci are easily distinguished from the background of fibroblasts. Foci consist of clusters of rounded refractile cells, each looking like a mitotic cell. However, all the cells in a focus retain their altered morphology throughout interphase. A focus often becomes multilayered (Figure 4.11B) as the rounded cells migrate on top of the fibroblasts in the surrounding monolayer. This migration may cause some of the transformed cells in the focus to detach, leaving a hole surrounded by the round refractile cells (Figure 4.11C). Such migration in culture is reminiscent of the metastasis of cancer cells in the body. Metastatic cells also detach readily from their site of origin and travel to new sites in the body, where they establish secondary tumors.

The RSV-transformed chick cells had alterations in their morphology and growth properties that made them appear tumorigenic, especially since these changes were promoted by a virus known to be capable of causing tumors in chickens. Nevertheless, subsequent research demonstrated that these transformed cells were not immortal. Like normal chick cells, the RSV-transformed cells could be passaged about 20 times, but then the cultures underwent *crisis* and the entire population died. Evidently this type of transformation is not sufficient to permit the cells to become immortal. We shall come back to the requirements for cell immortality later.

Another common assay for transformed cells takes advantage of their loss of an anchorage, or attachment, dependency. That is, they are able to grow without attaching to a tissue culture dish. Instead, some transformed cells are able to form large spherical colonies when they are grown in a medium containing soft agar. Ian Macpherson and Luc Montagnier demonstrated this ability in the early 1960s (Figure 4.12). Montagnier may be familiar to you as the French discoverer of the virus that causes AIDS. Figure 4.12 shows a colony of hamster cells from a line, BHK21/13, growing in agar seven days after infection by a DNA tumor virus called polyoma. The arrows point to single cells that did not divide, indicating that they were either not infected or not transformed.

Cellular transformation is not caused only by viruses. Betsy Sutherland and her colleagues transformed human cells by exposing cultures of human embryonic skin and muscle (HESM) fibroblasts to several short doses of ultraviolet light. Figure 4.13B shows a colony of HESM cells growing in soft agar 14 days after the irradiation. Unirradiated controls failed to grow under these conditions (Figure 4.13A). The experimenters report that these transformed human cells exhibited an extended life span, but some of the cultures derived from the anchorage-independent clones had a finite life span, and the cells were apparently nontumorigenic.

As we have seen, lymphocytes in culture can be stimulated by mitogens to divide at least several times. It is also known that these cultures do not form permanent lines under such conditions. However, when lymphocytes are obtained from patients who are suffering from acute infectious mononucleosis, or even from those who no longer show the acute symptoms of this disease, permanent cell lines can be readily

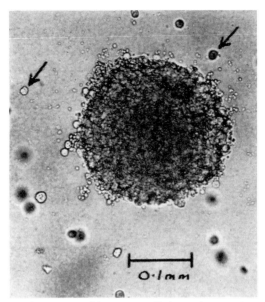

Figure 4.12 Transformation of hamster cells by polyoma virus.
A colony of transformed cells in agar medium seven days after the culture was
seeded with hamster cells (BHK21/13) that were infected with polyoma virus.
Single nondividing cells are indicated by the arrows.

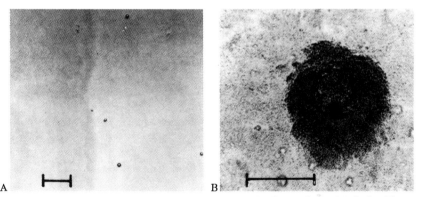

Figure 4.13 Transformation of human cells with ultraviolet (UV) light. The
scale shows 0.2 mm.
(A) Unirradiated human embryonic skin and muscle (HESM) fibroblasts form
no colonies 14 days after being placed in soft agar. (B) UV-irradiated HESM
cells form anchorage-independent colonies such as the one shown within 14
days after plating in soft agar.

established. Infectious mononucleosis, characterized by the appearance of enlarged tender lymph nodes, an enlarged spleen, and abnormal lymphocytes in the blood, is caused by a member of the herpes virus group called Epstein-Barr virus (EBV). This virus can transform and immortalize the B lymphocytes (the type that produce antibodies) in humans and a few other primate species.

Another type of immortalized lymphocyte can be obtained from lymphomas, or cancerous lymphocytes. A very common lymphoma prevalent in children in certain regions of Africa is called Burkitt's lymphoma, after the English physician who first studied it. This cancer of the B lymphocytes gives rise to extremely large tumors (Figure 4.14), and it is almost certainly caused by EBV. In Figure 4.15A we can see an electron micrograph of EBV developing within the nuclei of cells cultured from Burkitt's lymphoma. In Figure 4.15B a mature EB viral particle (virion), with its membranous envelope, capsid (protein coat), and nucleoid (internal DNA-protein complex), can be seen in the cytoplasm.

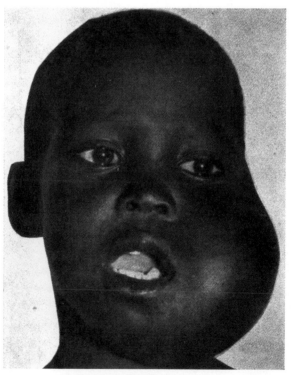

Figure 4.14 Burkitt's lymphoma.
This example involves the mandible of a 5-year-old boy.

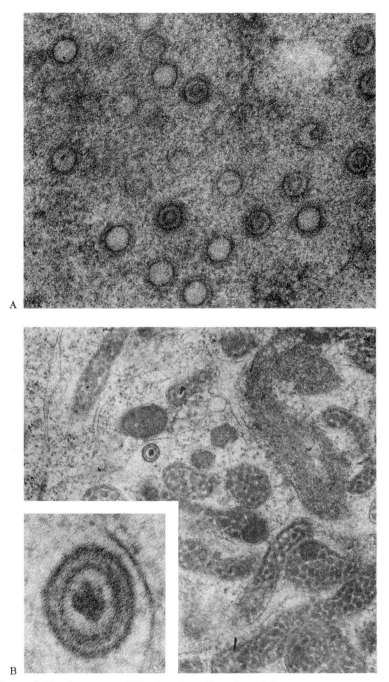

Figure 4.15   Structure of EB virus in cells cultured from Burkitt's lymphoma.
(A) Numerous developing immature particles in a thin section of a
lymphoblast nucleus (76,500✕). (B) Mature viral particle with envelope,
capsid, and nucleoid in the cytoplasm (42,000✕; *inset* 213,500✕).

4.2.3 Selection for Tumorigenic Cells

Clearly the tumorigenicity of a cell cannot be determined simply by the alteration of its morphology or growth properties, even when this transformation results from infection by a virus. One of the best assays to determine whether or not a particular type of cell is tumorigenic relies on the ability of the cells to grow into a tumor in a nude mouse. Nude mice, you remember, are mutants that fail to develop a thymus gland. For this reason, they have almost no immune defense system and fail to reject cells that are transplanted from a different species.

Charles Stiles and his colleagues demonstrated that although certain human cell lines of neoplastic (cancerous) origin formed tumors in nude mice (Figure 4.16A), human cell lines transformed by simian virus 40 (SV40) failed to do so when they were injected into these animals,

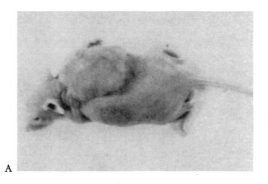

A

B

Figure 4.16 Appearance of human tumors and abortive nodules in nude mice. (A) Nude mouse with tumor formed from injection of WI-L2 cells photographed six weeks after injection. (B) Nude mice with abortive nodules induced by human SV40 transformants (the SV80 line). The animals were photographed one week after injection of 2×10^6 cells subcutaneously. The upper mouse was untreated and the lower mouse was treated with antilymphocyte serum. Arrows indicate the transient swellings (abortive nodules) that appeared at the site of injection.

and only transient swellings occurred at the sites of injection (Figure 4.16B). The investigators concluded that even immortal, virally transformed cell lines are not necessarily tumorigenic in animals.

4.2.4 Chromosomal Changes Upon Growth of Cultured Cells

Considerable effort has been expended to gain an understanding of the molecular and genetic basis of the relation between cell transformation and tumorigenicity. A very important approach to this problem is the study of hybrids that result from fusions between tumorigenic and normal cells. In most reported cases, the hybrids continue to express many traits characteristic of transformed cells, yet they do not produce tumors. One likely interpretation of this finding is that the normal cell possesses one or more genes that have the power to suppress the ability of even a cancer cell to express its tumorigenic properties.

Using hybrids created from fusions between human tumorigenic cells (HeLa cell line) and normal human fibroblasts, Eric Stanbridge and his colleagues demonstrated that the loss of certain specific chromosomes was apparently related to the conversion of these hybrids from the nontumorigenic to the tumorigenic state. Analysis of the chromosomes (*karyology*) in the parental and hybrid cell lines was carried out by several staining techniques, which enabled the researchers to detect the presence or absence of the Y chromosome, as well as all the other chromosomes and chromosomal rearrangements. They analyzed at least 25 metaphase spreads for each hybrid cell line over a period of at least 40 population doublings. Tumorigenicity of the hybrids in nude mice was tested simultaneously.

Typical karyograms of the two parental cell lines, HeLa variant D98/AH2 and a human *diploid* fibroblast (a cell that has two sets of chromosomes, one from each parent), are shown in Figure 4.17. The bizarre chromosomal pattern of the HeLa cells is typical of tumorigenic cells that have been passaged for many years. Such cells are said to be *aneuploid* (without the normal diploid number of chromosomes). The karyograms of four hybrid cell lines are seen in Figure 4.18. Lines CGL-1 and CGL-2 are nontumorigenic, whereas the other two (CGL-3 and CGL-4) are tumorigenic segregants. After a detailed statistical evaluation of the karyotypes of the four hybrid cell lines, Stanbridge and his associates concluded that the loss of one copy each of chromosomes 11 and 14 from the hybrid cells is correlated with the expression of tumorigenicity. Presumably, genes associated with these chromosomes also suppress tumorigenicity in normal human diploid cells.

It is also now well known that every human cell contains oncogenes, which can convert a normal cell into a tumorigenic one. The abnormal regulation of oncogene expression triggers this conversion.

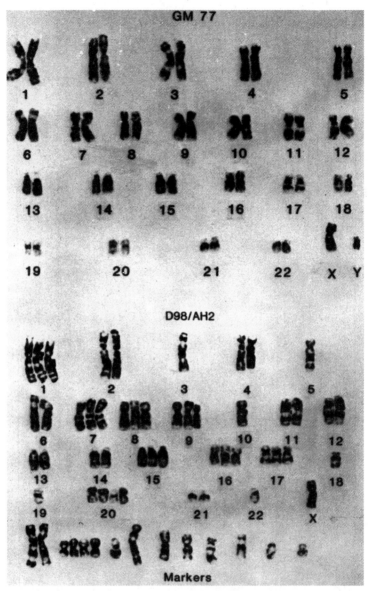

Figure 4.17 Representative karyograms of the two human parental cell lines: HeLa variant D98/AH-2 and human fibroblasts GM77.

Marguerite Vogt and her collaborators used murine (mouse) nonadherent spleen cells in their efforts to determine how oncogenes enable cells to survive the aging crisis and become immortal. From three weeks to several months in culture, these cells attained a steady state of growth (no increase in cell number) as a result of an equilibrium involving proliferation, terminal differentiation, and cell death. After three weeks

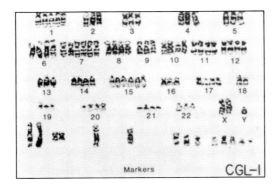

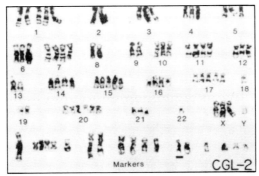

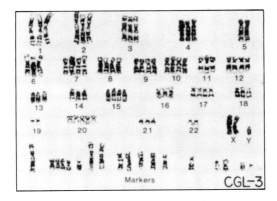

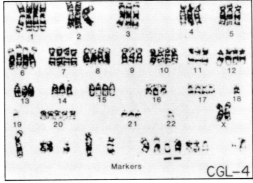

Figure 4.18 Representative karyograms of two nontumorigenic hybrid cell lines, CGL1 and CGL2, and two tumorigenic segregants, CGL3 and CGL4. HeLa markers are indicated and new hybrid-specific markers are underlined.

in culture, spleen cells were infected with various retroviruses carrying the oncogenes called *ras* and *myc*. Three months later, several significant changes were observed in the cultures infected with both *ras* and *myc* viruses. The incidence of cell death rose to 90 to 95 percent, and soon thereafter rapidly dividing heterogeneous cells emerged. "The changes in cultures," the experimenters reported, "were reminiscent of the changes observed when normal cells with a limited life span go into crisis and cells with an unlimited growth potential emerge" (Vogt et al., 1986, p. 3545). Most important, the cells that survived this apparent crisis grew in the absence of previously required growth factors (see Figure 4.19). This property is characteristic of both transformed and tumorigenic cells.

Vogt and her colleagues then went on to show that the growth-factor-independent cells exhibited gross changes in chromosome numbers around the time the cells passed through their aging crisis (between 75 and 103 days). In cultures infected with virus containing either the *ras* or the *myc* oncogene, but not both, nearly all of the cells (92 to 98 percent) retained their normal 40 chromosomes during the four months of analysis. In contrast, 27 percent of the cells that were infected with viruses containing both the *ras* and the *myc* oncogenes had hypodiploid chromosome numbers (34 to 39) during the first two to three months after infection (Figures 4.19 A–C). Soon after cell crisis, 85 percent of the cells from a single clone had karyotypes with chromosome numbers primarily around the tetraploid number (Figure 4.19D). The karyotypes of growth-factor-independent cells from single colonies varied greatly from colony to colony (Figures 4.19E and F). Even different cells from a single colony had chromosome numbers that varied from near haploid (21 chromosomes) to near triploid (64 chromosomes; Figure 4.20).

The investigators postulated that "unbalanced genomes (i.e., not diploid) caused by an abnormal distribution of chromosomes to the daughter cells after crisis may allow the expression of a recessive trait" (Vogt et al., 1986, p. 3545), thereby permitting the appearance of a tumorigenic phenotype; namely, growth-factor independence. The initiation of these changes in chromosome number is in some way caused by the expression of a combination of certain oncogenes such as *ras* and *myc*. Without the chromosomal changes that result in the loss of tumor-suppressor genes, a normal cell may never become fully trans-

---

Figure 4.19 Chromosome distribution in doubly infected (*ras* and *myc* oncogenes) spleen-derived monocytes at various times postinfection (p.i.). (A and B) Mass culture before crisis. (C) Colony #4 selected from mass culture, analyzed before crisis. (D) Colony #4 analyzed shortly after crisis. (E and F) Cells from growth-factor-independent colonies derived from colony #4, analyzed after crisis. The percentage of colony-forming cells in a medium lacking growth factors is indicated in each panel.

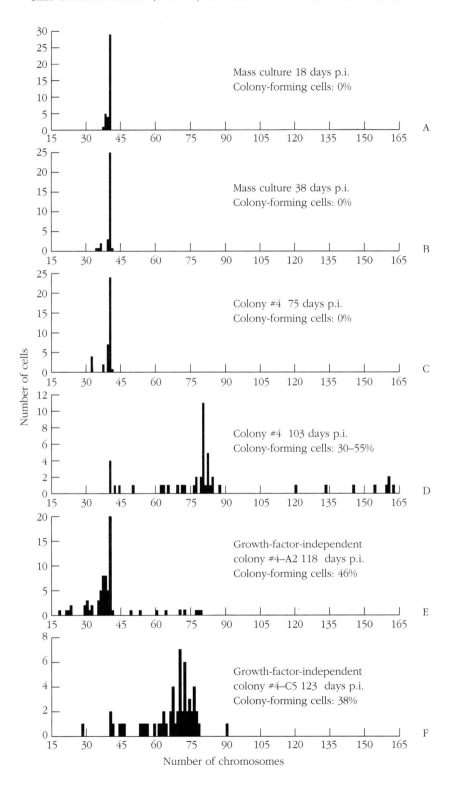

Mass culture 18 days p.i.
Colony-forming cells: 0%

A

Mass culture 38 days p.i.
Colony-forming cells: 0%

B

Colony #4 75 days p.i.
Colony-forming cells: 0%

C

Colony #4 103 days p.i.
Colony-forming cells: 30–55%

D

Growth-factor-independent
colony #4–A2 118 days p.i.
Colony-forming cells: 46%

E

Growth-factor-independent
colony #4–C5 123 days p.i.
Colony-forming cells: 38%

F

Number of cells

Number of chromosomes

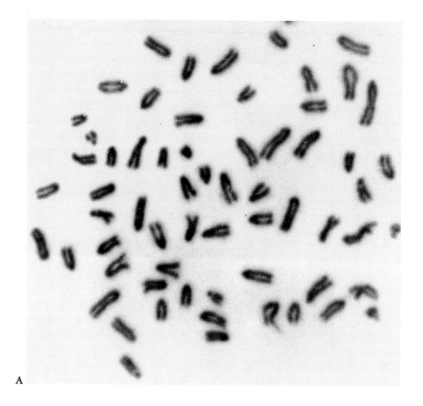

A

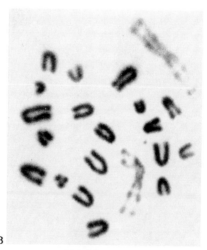

B

Figure 4.20 Chromosome spreads from two cells of colony 4-A2 (see Figure 4.19E).
(A) Near-triploid cell with 64 chromosomes. (B) Near-haploid cell with 21 chromosomes.

formed into a tumorigenic cell even if its cancer-causing oncogenes have been activated.

Given the enormous number of cells in the human body, the existence of over 40 oncogenes in each cell, and the fact that a tumor can originate in a single cell, cancer is apparently a relatively rare event involving a complex relationship among cell differentiation, survival of an aging crisis, and the expression of certain oncogenes and tumor-suppressor genes. Cell-culture studies combined with the powerful techniques of molecular biology are providing many rewarding insights into the mechanisms of these phenomena.

Discussion Questions

1. How were the existence of NPF and NTF first demonstrated to be related to the differentiation of nerve cells?

2. What evidence supports the view that myotube formation precedes the induction of muscle-specific enzymes?

3. Draw a diagram outlining the protocol of the Bender and Prescott experiment.

4. In Bender and Prescott's experiment, what percentage of the chromatids would be labeled (*a*) at the first metaphase after DNA labeling? (*b*) at the third metaphase after DNA labeling? (Do not confuse chromatids with chromosomes.)

5. What are two assays for determining the percentage of transformed cells present in a population growing in culture?

6. What property of nude mice enables them to support the growth of human tumor cells?

7. How were the techniques of cell fusion and karyotypic analysis used to demonstrate the existence of tumor-suppressor genes?

8. Identify three changes that occur in cells that have survived the crisis phenomenon.

9. Cite an example of an immortal cell line that is not tumorigenic. What genetic changes may be necessary for that line to become tumorigenic?

References

Bender, M. A., and Prescott, D. M. (1962). DNA synthesis and mitosis in cultures of human peripheral leukocytes. *Experimental Cell Research, 27*:221.

Epstein, M. A., G. Henle, B. G. Achong, and Y. M. Barr (1965). Morphological and biological studies on a virus in cultured lymphoblasts from Burkitt's lymphoma. *Journal of Experimental Medicine, 121*:761.

Ham, R. G. (1965). Clonal growth of mammalian cells in a chemically defined, synthetic medium. *Proceedings of the National Academy of Sciences USA, 53*:288.

Harris, H., and J. F. Watkins (1965). Hybrid cells derived from mouse and man: artificial heterokaryons of mammalian cells from different species. *Nature, 205*:640.

Hayflick, L. (1965). The limited *in vitro* lifetime of human diploid cell strains. *Experimental Cell Research, 37*:614.

Herzenberg, L. A., R. G. Sweet, and L. A. Herzenberg (1976). Fluorescence-activated cell sorting. *Scientific American, 234*:108.

Illmensee, K., and B. Mintz (1976). Totipotency and normal differentiation of single teratocarcinoma cells cloned by injection into blastocysts. *Proceedings of the National Academy of Sciences USA, 73*:549.

Macpherson, I., and L. Montagnier (1964). Agar suspension culture for the selective assay of cells transformed by polyoma virus. *Virology, 23*:291.

Manthorpe, M., S. Skaper, and S. Varon (1980). Purification of mouse Schwann cells using neurite-induced proliferation in serum-free monolayer culture. *Brain Research, 196*:467.

Martin, G. M., C. A. Sprague, and C. J. Epstein (1970). Replicative life-span of cultivated human cells: effects of donor's age, tissue, and genotype. *Laboratory Investigations, 23*:86.

Milstein, C. (1980). Monoclonal antibodies. *Scientific American, 243*:66.

Pollack, R. (ed.) (1975). *Readings in Mammalian Cell Culture.* Cold Spring Harbor Laboratory of Quantitative Biology.

Richler, C., and D. Yaffe (1970). The *in vitro* cultivation and differentiation capacities of myogenic cell lines. *Developmental Biology, 23*:1.

Ruddle, F. H. (1982). A new era in mammalian gene mapping: somatic cell genetics and recombinant DNA methodologies. *Nature, 294*:115.

Shainberg, A., G. Yagil, and D. Yaffe (1971). Alterations of enzymatic activities during muscle differentiation *in vitro. Developmental Biology, 25*:1.

Stanbridge, E. J., R. R. Flandermeyer, D. W. Daniels, and W. A. Nelson-Rees (1981). Specific chromosome loss associated with the expression of tumorigenicity in human cell hybrids. *Somatic Cell Genetics, 7*:699.

Stiles, C. D., W. Desmond, Jr., G. Sato, and M. H. Saier (1975). Failure of human cells transformed by simian virus 40 to form tumors in athymic nude mice. *Proceedings of the National Academy of Sciences USA, 72*:4971.

Sutherland, B. M., J. S. Cimino, N. Delihas, A. G. Shih, and R. P. Oliver (1980). Ultraviolet light-induced transformation of human cells to anchorage-independent growth. *Cancer Research, 40*:1934.

Temin, H., and H. Rubin (1958). Characteristics of an assay for Rous sarcoma virus and Rous sarcoma cells in tissue culture. *Virology, 6*:669.

Varmus, H. E. (1984). The molecular genetics of cellular oncogenes. *Annual Review of Genetics, 18*:553.

Varmus, H., and A. J. Levine (eds.) (1983). *Readings in Tumor Virology.* Cold Spring Harbor Laboratory of Quantitative Biology.

Varon, S., M. Manthorpe, L. R. Williams, and F. H. Gage (1988). Neurotrophic factors and their involvements in the adult CNS. In *Aging and the Brain* (R. Terry, ed.), p. 259. New York: Raven Press.

Vogt, M., J. Lesley, J. Bogenberger, S. Volkman, and M. Haas (1986). Coinfection with viruses carrying the v-Ha-*ras* and v-*myc* oncogenes leads to growth factor independence by an indirect mechanism. *Molecular Cell Biology, 6*:3545.

Weinberg, R. A. (1983). A molecular basis of cancer. *Scientific American, 249*:126.

Yelton, D. E., and M. D. Scharff (1981). Monoclonal antibodies: a powerful new tool in biology and medicine. *Annual Review of Biochemistry, 50*:657.

5

Genetic Engineering: Techniques and Applications

The goals of genetic engineering have been a major focus of biological research for centuries. Genes of all sorts have been manipulated in plants, animals, and microorganisms by botanists, horticulturists, geneticists, industrial and medical microbiologists, and more. Whereas their techniques relied on mutation and selection, in accordance with the laws of classical genetics, today's genetic engineers use a new set of biochemical methods for transferring specific genes into the organisms of choice. The end result remains the same: the creation of genetically modified organisms that can benefit humankind.

Modern genetic engineering is both simple and ingenious. With this technology we can do such wonderful things as clone human genes that determine the synthesis of insulin and detect genetic diseases before a child is born. We can introduce human genes into mice and then study the mice to learn much that we don't yet know about the basis and treatment of human diseases. The power of these recently developed techniques is far-reaching, and their applications are expected to develop quickly into a multibillion-dollar biotechnology industry.

5.1 Genetic Tools

When we watch any drama, we have to be introduced to the characters before the plot can thicken. Our cast is composed of two important families, the plasmids and the restriction endonucleases. Each family has many members, but we need to know only a few of them to grasp the essential principles of genetic engineering.

5.1.1 Plasmids

Plasmids are relatively small rings of DNA that, like viruses, replicate independently inside a bacterial cell. They are not essential to the bacteria. Instead they have a symbiotic relationship with their hosts, providing them with resistance to such antibiotics as penicillin or with some other valuable trait. The bacteria in turn provide plasmids with the machinery to replicate and synthesize proteins. Plasmids are generally present in multiple copies within a bacterial cell, a property that results from their genetic ability to replicate independently of the DNA of the host cell, much like viruses. Different plasmids maintain from 3 to 50 copies of DNA per cell. As many as 3000 copies per cell can be generated under certain experimental conditions. These characteristics make plasmids an obvious choice for use as genetic vehicles (*vectors*) for introducing multiple copies of a selected gene into a cell.

5.1.2 Restriction Endonucleases

Bacteria possess a wide assortment of enzymes called *restriction endonucleases*, which cut foreign DNA into pieces. Their own DNA is protected from such cleavage by methylation of nucleotide bases at sites recognized by the restriction enzymes. There are two basic types of restriction endonucleases. Those of type I bind to a specific DNA sequence and cleave at sites far removed from it. Type II enzymes also recognize specific sequences, but cut at specific sites within those sequences. Type II enzymes are used in *recombinant DNA* technology because one knows exactly where they will cut a given DNA molecule. Sequences recognized by type II restriction endonucleases are generally from four to six base pairs long and are palindromes. *Palindromes* are words or sequences of symbols that read the same forward and backward, such as *radar*. In a palindromic sequence of DNA, the reading of one strand from left to right results in the same sequence found in the other strand when read from right to left (see Figure 5.1). The restriction enzymes create two types of double-stranded breaks in DNA. One type results in blunt ends that have no unpaired bases, and the other more

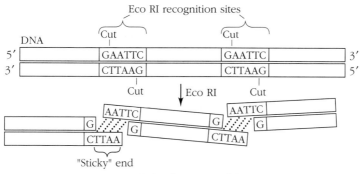

Eco RI recognition sites

Restriction fragments of DNA

Figure 5.1 Cutting of DNA by restriction enzyme Eco RI.
The endonuclease Eco RI recognizes the indicated six-base-pair sequence in
DNA and makes staggered cuts in both strands of the DNA molecule. This
leaves fragments of DNA with single-stranded ("sticky") ends consisting of
four bases. Note that all Eco RI fragments can form hydrogen bonds with each
other via their "sticky" ends, either in their original combination or in new,
recombinant combinations.

common type creates homologous single-stranded "sticky" ends, as
shown in Figure 5.1.

Restriction endonucleases are remarkable tools in recombinant
DNA technology. With them one can cut at specific sequences in a
plasmid or any other DNA, leaving homologous ends that can readily
rejoin. To identify these DNA fragments, gel electrophoresis is most
frequently the method of choice (Figures 5.2 and 5.3). Since gel electrophoresis separates molecules according to size (as we saw in Chapter 2),
one can separate various DNA fragments formed by restriction endonucleases. A gel resulting from the digestion of DNA from SV40 virus with
Hind III endonuclease is seen in Figure 5.2. Lane 1 contains undigested
DNA, and lane 2 has the digested DNA. The map and length of the
fragments are shown above the photographs of the gels. Note that the
rate of movement through the gel is inversely related to the size of
the DNA fragment.

The DNA fragments themselves can be viewed when the gel is
immersed in an ethidium bromide solution (Figure 5.3). The ethidium
bromide binds to the DNA between the base pairs and causes the DNA to
fluoresce under ultraviolet light. Another way to determine the location
of the DNA fragments is by the Southern blot method (see Figure 5.4).
After gel electrophoresis of the fragments, the DNA is denatured and
transferred from the gel by capillary action onto a sheet of nitrocellulose. Then a solution containing the desired radioactively labeled nucleic acid sequence (a *probe*) is incubated with the nitrocellulose. The

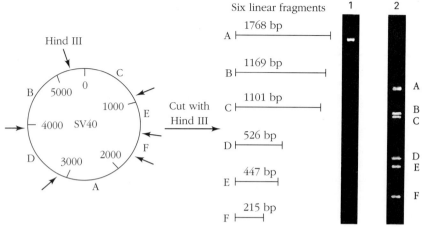

Figure 5.2 Hind III map of SV40.
The DNA from SV40 virus can be purified and digested with the restriction endonuclease Hind III (from *Haemophilus influenzae*). Both the digested and the undigested DNA can then be subjected to electrophoresis. The DNA becomes visible in the gel by incorporating ethidium bromide, a molecule that binds to DNA and fluoresces under ultraviolet irradiation. Lane 1 represents the uncut DNA and lane 2 the digested DNA. Hind III cuts the SV40 molecule six times, producing six fragments. By convention, the pieces of DNA released by a restriction enzyme are labeled A to Z in order of decreasing size; the Hind III fragments of SV40 are therefore labeled A to F. The sizes of the various pieces are given in the diagram (base pairs, bp).

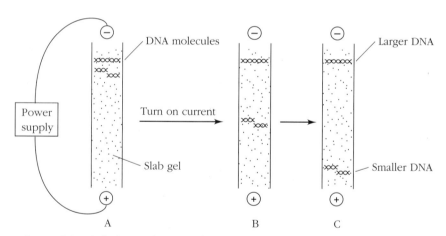

Figure 5.3 Gel electrophoresis of DNA.
(A) Gel electrophoresis separates macromolecules on the basis of their rate of travel through a gel in an electrical field. Negatively charged DNA molecules are introduced near the negatively charged end of a thin slab of a polymeric gel — such as polyacrylamide or agarose — that is supported by glass plates and bathed in an aqueous solution. Electrodes are attached to both ends and the current is turned on. (B) DNA molecules migrate toward the positive pole at a rate determined primarily by their size. Larger DNA molecules move more slowly because they have more difficulty fitting through the spaces in the gel. (C) After sufficient time of electrophoresis, the DNA molecules of various sizes separate into distinct bands in the gel. The bands can be made visible by a fluorescent dye — such as ethidium bromide — that binds to the DNA in the gel. Under ultraviolet light, the DNA bands fluoresce pink.

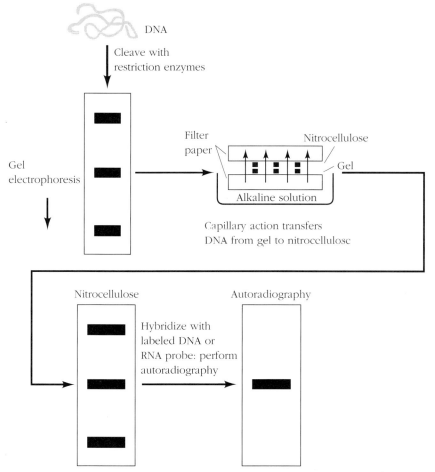

Figure 5.4   The Southern blot technique for detecting the presence of specific DNA sequences.

DNA digested with one or more restriction enzymes is separated into fragments by gel electrophoresis. Marker DNA fragments of known lengths are included in the gel so that their positions can be used to estimate the lengths of the experimental fragments. The DNA is denatured and transferred from the gel to a nitrocellulose sheet by capillary action. Specific labeled sequences of DNA or RNA can then be hybridized to the bound DNA. Autoradiography locates the hybridized nucleic acids, which indicate the presence of particular known sequences in specific restriction fragments.

labeled probe hybridizes with the homologous DNA molecules by forming hydrogen bonds. Finally, the nitrocellulose is dried and subjected to autoradiography. The DNA complementary to the probe shows up as a dark band on the film because the radiation exposes the film at that spot. This technique is sensitive enough to detect one part in a million of DNA with as little as 5 μg of a DNA sample.

5.1.3 Plasmids and Restriction Enzymes United

The first person to work on recombining DNA from different species was Paul Berg. In 1971 he took DNA from an animal virus, SV40, and combined it with DNA from a bacteriophage, a virus that attacks only bacteria. Because he did not have restriction enzymes that generated sticky ends, he added on complementary tails to the ends of the DNA fragments so that they would recombine in vitro. It would have been easy for him to get this recombinant DNA into bacteria, but he had ethical reservations about developing this new technology before adequate discussion of its potential dangers.

One year later, in November 1972, Herbert Boyer and Stanley Cohen met at a U.S.A./Japan conference on bacterial plasmids in Honolulu, Hawaii. Upon hearing each other's presentations, they immediately realized the possibility of replicating genes that had been inserted into a new host. First they needed a plasmid that would be cut only once by a restriction enzyme such as Eco RI while not losing its potential for replication. It also had to have at least one antibiotic-resistance gene left intact so that they could select for a clone possessing the desired gene. *Antibiotic-resistance genes* enable bacteria to synthesize a protein that either inactivates the antibiotic or prevents it from entering the cell. Penicillinase, for example, is an enzyme that destroys the activity of penicillin, making bacteria that have this genetic capacity resistant to that particular antibiotic.

A suitable plasmid was found among the many in Cohen's collection at Stanford. It was named pSC101 and carried a gene that confers resistance to the antibiotic tetracycline (Tet^R). This plasmid was then combined with a kanamycin-resistance ($Kana^R$) gene that had been fragmented off from the rest of its plasmid by Eco RI. Figure 5.5 is a diagram of the process of gene cloning in a plasmid.

Figure 5.5 Using plasmids for cloning genes.
(A) Plasmids possess genes which make the bacteria that harbor them resistant to certain antibiotics, such as ampicillin and tetracycline (the Amp^R and Tet^R genes). Plasmid DNA is purified from the bacteria. (B) Foreign DNA that contains the desired gene is purified from cells. (C) One restriction enzyme cuts the plasmid DNA and the foreign DNA at the same nucleotide sequence, thereby generating complementary "sticky" ends. (D) The plasmid DNA and the foreign DNA are incubated together in the presence of DNA ligase, an enzyme that catalyzes the formation of covalent bonds between DNA fragments. A "recombinant" plasmid is thereby created. (E) The recombinant plasmid enters a bacterial cell by a procedure called "transformation." Colonies of bacteria carrying the desired gene are selected by growing the transformed cells in agar containing the appropriate antibiotics. Since the foreign gene was inserted into the plasmid by a cut in the Tet^R gene, this gene becomes inactive, and the cells will grow on agar containing ampicillin but not tetracycline.

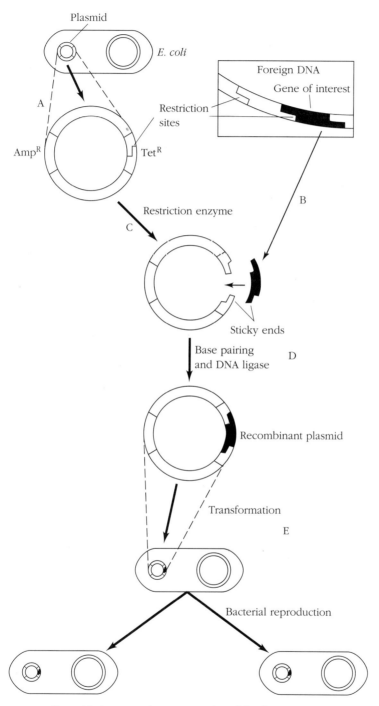

Bacterial clone carrying many copies of the foreign gene

Cohen and Boyer used Tet^R as the plasmid marker gene and Kana^R as the inserted gene of interest. Since both genes resided on DNA fragments that had been cut by Eco RI, they had complementary sticky ends that could join by hydrogen bonding. Cohen and Boyer incubated the plasmid with its newly added Kana^R gene with T4 DNA ligase, an enzyme that covalently links the DNA fragments, thereby completing the formation of an intact circular DNA molecule. Finally, they transformed the bacteria with the recombinant plasmid, a process that requires that the bacterial membrane be made permeable so that the plasmid DNA can be taken into the bacterial cells. Cohen and Boyer used media with both tetracycline and kanamycin to kill off the bacteria that did not possess the hybrid plasmid, so that only those bacteria that contained *both* antibiotic-resistance genes formed a *bacterial colony.* By this means they cloned the newly inserted Kana^R gene.

Cohen and Boyer and their colleagues completed the splicing of foreign genes into plasmids and the introduction of these newly designed plasmids into bacteria by March 1973. The original experiments involved genes and plasmids derived from a single bacterial species, *E. coli.* Within a year, however, these scientists broke the species barrier to the exchange of genetic material. Genes from a eukaryotic organism, a frog known formally as *Xenopus laevis,* were introduced into *E. coli* via a plasmid and copied efficiently into RNA. Incredibly rapid progress in this exciting new field of recombinant DNA technology followed.

5.2   Important Techniques for Cloning Genes

Now that we have these remarkable tools called plasmids and restriction endonucleases, it is instructive to explore some of the various techniques for cloning particular genes. Though the concepts are easy to understand, many subtleties make this endeavor a real challenge and leave room for considerable creativity.

5.2.1   Necessary Qualities in a Vector

The selection of a vector is critical in cloning. The decision affects how one will *transfect* the bacteria (that is, introduce the foreign DNA into its new host), how many copies of the gene will be made, how easily the transfected bacterial clone can be selected, and the efficiency of expression of the gene product. Two types of vectors are commonly used: viruses and plasmids. The viral form is often bacteriophage lambda, mutated in such a way that it will not kill the infected cell. Phages provide a better means of transfecting bacteria because they inject their DNA into the recipient cells efficiently. Methods for detecting a transfected bacterial clone and the expression of a gene are much the same for plasmids and phages.

If a plasmid is to be a suitable vector, it must first possess a site of replication so that it can replicate independently of its host cell. One commonly used plasmid, pBR322, replicates very quickly, generating from 50 to 400 copies per bacterial cell. Being small, this plasmid is relatively easy to sequence and analyze. Second, the plasmid should possess only one recognition site for the restriction endonuclease being used, thus ensuring the retention of the entire plasmid in the desired recombinant. For efficient expression of a gene, a bacterial promoter is usually inserted next to the desired gene (see Figure 5.6). *RNA polymerase* recognizes the nucleotide sequence in the promoter and, upon binding, unwinds the DNA and begins the process of transcription. Transcription stops when the RNA polymerase reaches a *terminator sequence* on the DNA. Finally, the plasmid must have a ribosome binding site, a sequence that initiates translation and controls its rate. Thus several types of control sequences in DNA help determine the efficiency of gene expression.

Genetic markers are also helpful in identifying the colony containing recombinant DNA. Recall that Cohen and Boyer designed their plasmid to code for resistance to kanamycin and tetracycline. Such antibiotic-resistance genes made it easy for them to select for colonies containing the recombinant DNA. Any colonies that could not grow on both tetracycline and kanamycin did not contain their newly inserted gene. The plasmid can also be manipulated to select for colonies that

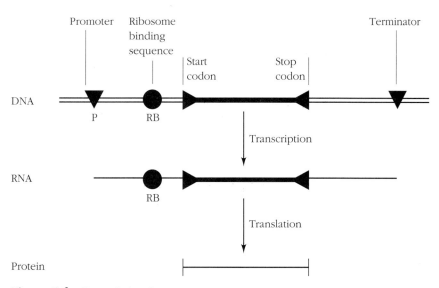

Figure 5.6   Control signals in gene expression.

have recombined with foreign DNA. When the DNA is inserted into one of the two antibiotic-resistance genes—for example, Tet^R—any colonies that contain a recombinant plasmid will then be resistant to the other antibiotic but sensitive to tetracycline. Colonies having a plasmid that did not combine with foreign DNA would be resistant to both antibiotics. Another method is to splice the foreign gene onto the end of a bacterial gene that codes for an enzyme such as β galactosidase. The degree to which colonies grown with X-gal, a substrate for β galactosidase, turn blue indicates the presence and quantity of β galactosidase produced (see Section 5.3.2). Since the desired gene is attached to the one that codes for β galactosidase, it is probably being expressed along with that marker gene.

5.2.2 Obtaining the Gene

The several ways in which one can approach the cloning of a gene are summarized in Figure 5.7. The first technique, synthesizing genes, used to be looked upon with less favor than the other methods because it was quite tedious. Now it is much easier to synthesize genes, although still quite costly. If the amino acid sequence of a protein is known, the DNA sequences that code for it can be deduced from the genetic code. Alternatively, if the gene has been isolated, it can be sequenced in a relatively short time.

The second gene-cloning procedure involves the production of a *cDNA library*. In this method, one isolates all of the mRNA from cells that actively express the gene of choice by passing the total cellular RNA through a column with polythymidylic acid (poly T) residues bound to it. These residues bind the polyadenylic acid (poly A) sequences found at the 3′ end of almost all mRNAs. Recall that poly A sequences are not present on other types of RNA. Using gel electrophoresis or sedimentation in sucrose gradients, one then selects for mRNAs of a size close to that necessary to code for the desired protein. The more precisely one selects for the correct mRNA, the easier it becomes to obtain the clone containing the desired gene. After the mRNA is eluted from a gel or the proper fractions are pooled from a gradient, the mRNA is mixed with the enzyme reverse transcriptase and a primer for complementary DNA (cDNA) synthesis called oligodeoxythymidylic acid (oligo dT). The reaction generates cDNA copies of the mRNA, as diagrammed in Figure 5.8. After the RNA is removed by alkaline digestion or by enzymatic digestion with RNAase, the cDNA is left with a hairpin loop. The cDNA now serves as a primer for double-stranded DNA synthesis by means of DNA polymerase or reverse transcriptase. The final step in the process is the removal of the hairpin loop by digestion with S1 endonuclease, thereby generating a normal double-stranded DNA molecule. Since we started out with a collection of mRNAs of a particular size, we now have a

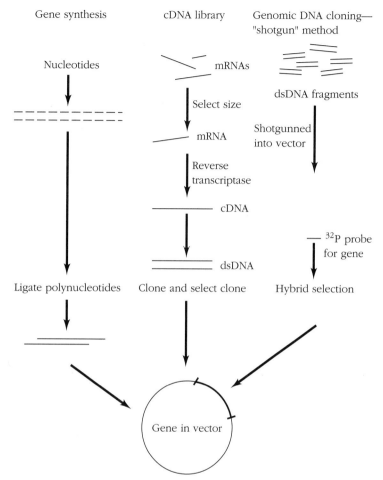

Figure 5.7 Methods to isolate desired genes.

set of DNAs coding for those mRNAs. All of these DNA molecules are then inserted into vectors with the use of restriction enzymes. This collection of recombinant plasmids is referred to as a cDNA library.

The third method shown in Figure 5.7, *genomic DNA cloning*, is also called the "shotgun" method. In this technique, the entire genome is fragmented and inserted into vectors. Cellular DNA is purified and chopped into double-stranded DNA fragments by restriction enzymes, such as Eco RI, which leave sticky ends. These fragments are then inserted into a vector that has the appropriate complementary sticky ends. Unlike the cDNA library, these genomic clones contain newly inserted DNA fragments, complete with introns and exons. Recall from Chapter 3 that introns are the noncoding regions interspersed between the exons, or coding regions, of nearly all eukaryotic genes. Depending

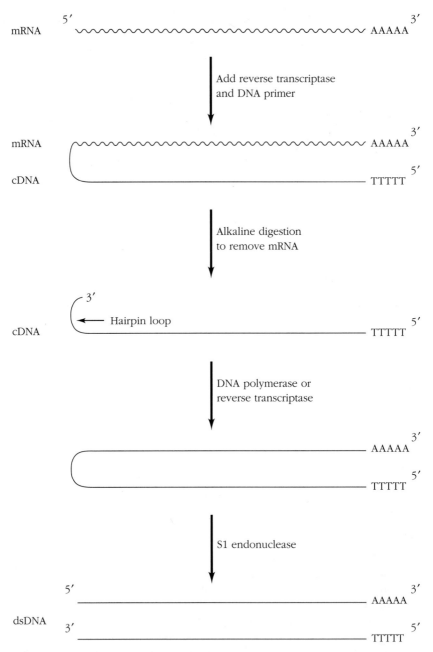

Figure 5.8 Production of double-stranded DNA (dsDNA) from mRNA.

on the sites of cutting by the restriction enzymes, the inserted DNAs may be longer or shorter than the length of a gene. Now let us investigate the techniques for identifying a particular gene.

5.2.3 Finding the Desired Clone

After clones are selected from bacterial colonies by antibiotic resistance or the color formed on X-gal medium (see Section 5.3.2), it becomes necessary to pick out one or more clones with the desired gene. One way to do this is by DNA-DNA or DNA-mRNA hybridization, a technique developed by Michael Grunstein and David Hogness in 1975. First one makes a replica of the bacterial colonies by pressing a sterile cloth on top of the original colonies, then pressing this cloth onto another petri plate containing agar and nutrients for bacterial growth (Figure 5.9). Then the cells in the colonies are lysed (broken open) by

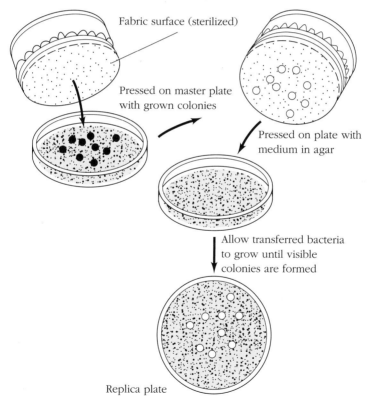

Fabric surface (sterilized)

Pressed on master plate with grown colonies

Pressed on plate with medium in agar

Allow transferred bacteria to grow until visible colonies are formed

Replica plate

Figure 5.9 Replica plating.
Colonies of bacteria are transferred from the surface of one agar plate to another by means of a sterile fabric pad. Some cells from each colony adhere to the fabric and then to the agar on the new plate. The cells quickly replicate and form colonies in the exact position on the new plate as on the original.

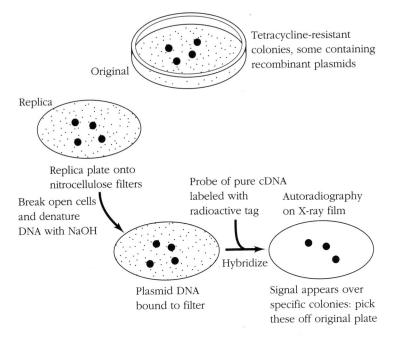

Figure 5.10 Detection of genes in colonies by DNA-DNA or DNA-RNA hybridization.
A radioactive DNA or RNA probe is used to detect the DNA of interest in a colony. This procedure is called "in situ" hybridization because the DNA is detected "in the place" that it is located.

the addition of alkali, which also separates the liberated DNA into single strands. A nitrocellulose filter disc is placed over the lysed colonies, and the single-stranded DNA adheres to it in a position corresponding to that of the original colony (Figure 5.10). Finally, a "hot probe" consisting of a radiolabeled DNA or mRNA specific for the desired gene is incubated with the nitrocellulose disc. The probe binds specifically to the DNA that came from colonies containing the desired gene. The radiolabeled filter is then placed on film that is sensitive to the emission of β particles from the hot probe. Thus an autoradiogram is produced, and colonies with the desired gene can be readily identified on the master plate (see also Figure 5.4).

A clever variation of this replica plating technique can be generally applied to the selection of colonies that produce any specific gene product for which an antibody is available. The use of this method, a modification of the so-called Western blot technique, to detect clones that produce insulin is diagrammed in Figure 5.11. Antibodies specific to insulin are first attached tightly to a plastic plate that fits perfectly into a petri dish. Insulin molecules that are being produced by the bacterial colonies can thus become attached to antibodies at corresponding sites on the plastic plate (step 2). The plate is then inserted into a solution of radioactively tagged insulin antibodies (step 3), which bind to the at-

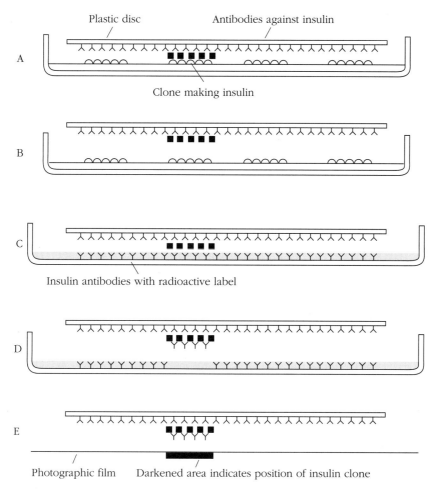

Figure 5.11 The "sandwich assay" for colonies producing a specific protein (insulin).
(A) Antibodies that bind to a specific protein (insulin in this example) are fixed tightly to a plastic plate. (B) When this plate is placed on the surface of an agar dish containing colonies, any insulin that is secreted by the colonies will attach to the antibiotics. (C) The plate with the attached insulin molecules is then placed into a solution containing radioactive antibodies that also bind specifically to insulin. (D) A "sandwich" is formed that contains antibodies, insulin, and radioactive antibodies all attached to a plate in the position of the original insulin-producing colony. (E) The plate is placed on a photographic film, which becomes darkened in the position of the radioactive antibodies. The original insulin-producing colony can then be identified.

tached insulin (step 4). The plastic plate with its sandwich of antibody, insulin, and radioactive antibody is then placed on a photographic film, which becomes darkened at the position of the radioactive antibodies (step 5). One can thus locate the colonies that produce insulin by placing the film (the autoradiogram) in the proper orientation over the original petri dish.

5.3 Gene Cloning and Diabetes

Now that the characters have been described and the stage has been set, we are nearly ready to unfold the dramatic story of the cloning of the human insulin gene. A detailed analysis of this classic example of genetic engineering ought to give you a better understanding and appreciation of this field of research. But before we get into the events that led to the cloning of the human gene that codes for insulin, we should examine some features of the structure and function of this essential hormone.

5.3.1 Insulin and Diabetes

Insulin is a two-chain peptide hormone produced and secreted by the B cells of the pancreas, cell clusters located in the islets of Langerhans and specialized for the production and secretion of insulin. Within minutes after the level of blood glucose increases, the B cells release insulin, thereby promoting an increased rate of glucose uptake by most cells in the body. If insulin remains in circulation for a prolonged period of time, the production of glycogen and fat increases.

Diabetics generally have a deficiency either in hormone production or in insulin receptor function. In both cases, blood glucose levels increase to concentrations far above normal. The concentration of glucose is so high that it must be excreted in the urine. To excrete such large quantities of glucose, the body must produce large amounts of urine, and so it depletes its store of fluids. The diabetic has symptoms of excessive thirst and urination and rapidly loses weight. Cells continually demand glucose, which is stored in the body as glycogen and fat. Since diabetics have a deficiency in insulin activity, they can neither store new glycogen and fats nor make efficient use of the glucose formed from catabolism of glycogen and fats. This malfunction leads to many severe complications, such as atherosclerosis, kidney failure, blindness, and the loss of limbs.

Over 90 percent of diabetics can be divided into two classes: type I (insulin-dependent) and type II (insulin-independent). Type I, or juvenile-onset diabetes, is the more serious of the two, and it accounts for 15 percent of the total number of diabetics. Insulin must be taken daily if this type of diabetes is to be controlled. The remainder of the diabetics (10 million in the United States), those who have type II or adult-onset diabetes, can usually control their disease with drugs or careful diet, but not with insulin. In fact, insulin levels are often normal in type II diabetics.

Although type I diabetics have a difficult life and a bleak prognosis, they are much better off today than diabetics who lived before 1921. It was at this time that Frederick Banting and Charles Best discovered that the treatment of diabetics with their newly discovered hormone called insulin caused dramatic improvements. The next step in understanding the mechanism of insulin's action was to determine its structure. Frederick Sanger accomplished a Herculean feat by elucidating the amino acid sequence of insulin in 1953, when most scientists did not even know that a given protein had a precisely defined amino acid sequence. With modern technology, protein sequencing is today a common, automated procedure.

Insulin is first produced as a larger polypeptide chain called preproinsulin (Figure 5.12). The first 23 amino acids in this molecule serve as a signal to direct it to *secretory vesicles* (granules) that are stored inside the B cells of the pancreas. This signal sequence is enzymatically removed in the vesicle, leaving proinsulin. When it is time for insulin to be secreted, specific peptidases cleave out a sequence of 34 amino acids from the middle of the proinsulin molecule. Dipeptides are then removed from the ends of this excised peptide, forming the C peptide. Now insulin is in its functionally active form, with two chains, A and B, held together by disulfide bridges. One reason for this elaborate process is that the C peptide causes the proinsulin molecule to assume a conformation that is required for the formation of the disulfide bonds.

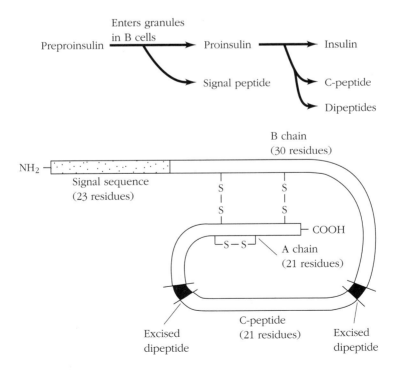

Figure 5.12 Steps in the production of insulin in vivo.

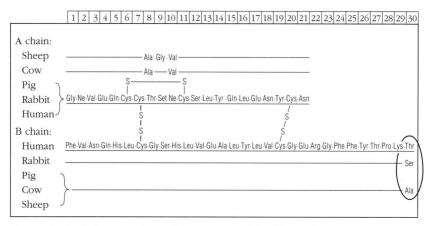

Figure 5.13   Primary amino acid sequence of insulin in five species.

In 1976 insulin was working miracles in type I diabetics, but it was not perfect. First of all, the only source of insulin was porcine (swine) and bovine (cow) pancreases. One could not obtain enough human pancreases to produce sufficient quantities of human insulin. One problem with the animal insulins is that they differ from human insulin in one or more amino acids (Figure 5.13). The differences are slight, but they cause allergic reactions in about 5 percent of diabetics. Some doctors suspected that certain complications resulting from diabetes, such as blindness, were caused by the animal insulins. Furthermore, animal pancreas supplies were decreasing, while the diabetic population was increasing at the rate of 5 percent a year.

5.3.2   Engineering a Clone for the Production of Human Insulin

Eli Lilly and Company, a major U.S. pharmaceutical firm, noted the gap between supply and demand in 1976 and recognized an opportunity to capitalize on emerging biotechnology. By means of recombinant DNA technology, they hoped to gain control of the market for insulin. Since this technology was relatively new, Eli Lilly was not equipped with the facilities and personnel to develop this product. So they did the next best thing—they convened a symposium to stimulate development of this project. Key figures in insulin research and recombinant DNA technology attended, enthusiastic about the possibilities of such a project. Thus began the race among various groups to clone the insulin gene and mass-produce its hormone product.

Three groups participated in the race to clone insulin: the Rutter-Goodman group, composed of researchers at the University of California at San Francisco (UCSF); the Gilbert group, based at Harvard and associated with a new biotechnology company, Biogen; and the UCSF – City of Hope group, which was affiliated with another biotechnology firm, Genentech. Genentech, in turn, was supported by Eli Lilly. Each of these teams contributed invaluable techniques to recombinant DNA technology within a few fast-paced years.

Cloning the Rat Insulin Gene

The first breakthrough, reported in June 1977, was made by the Rutter-Goodman group, who succeeded in cloning a rat insulin gene in *E. coli*. Although their clones did not secrete insulin, it now became clear that this mammalian gene could replicate under the control of bacterial plasmids. In addition, they purified sufficient quantities of rat insulin DNA to determine this gene's exact nucleotide sequence.

In their cloning procedure, Rutter and Goodman used one technique that has not yet been described: the addition of *synthetic linkers* (Figure 5.14). After obtaining the cDNA for insulin, they added oligonucleotides that contain Hind III restriction sites to both ends of the DNA by a technique called *blunt-end ligation* (step 4 in Figure 5.14). Relatively short oligonucleotides of specific sequence are easily synthesized and are now commercially available. They can be joined to the blunt ends of double-stranded DNA by means of T4 DNA ligase. Homologous single-stranded (sticky) ends were produced in the linkers and in the plasmid pBR322 by treatment with Hind III (step 5). Finally, the cDNA containing the rat insulin gene was inserted into the plasmid, the ends again joined with T4 ligase (step 6).

The Rutter-Goodman group was first to clone the insulin gene into a bacterium. Their results set the stage for the further work of getting expression of the rat insulin gene in bacteria and the final goal of producing human insulin with a bacterial clone.

Production of Rat Insulin by a Cloned Gene

Although Walter Gilbert's group did not win the race to clone the rat insulin gene, it was the first to clone a gene that produced insulin in bacteria. Furthermore, with a clever manipulation, the group even succeeded in getting the bacteria to secrete this hormone, thereby simplifying its purification in large-scale production.

The first step in the procedure, outlined in Figure 5.15, was the synthesis of cDNA coding for the entire preproinsulin molecule. A large quantity of mRNA was isolated from rat insulinoma cells grown in culture. These cells were derived from a pancreatic B-cell tumor that produces large amounts of insulin, and they thus contained a relatively high percentage of preproinsulin mRNA. Together with a synthetic primer,

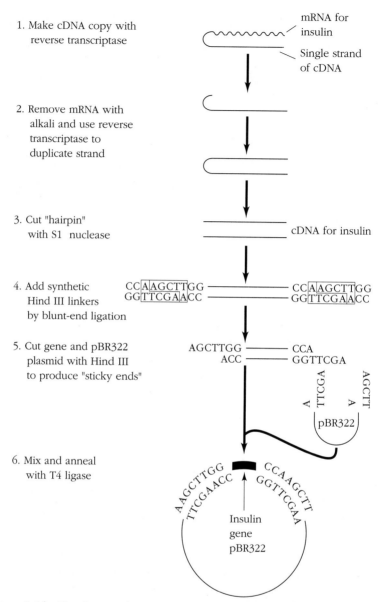

1. Make cDNA copy with reverse transcriptase

mRNA for insulin

Single strand of cDNA

2. Remove mRNA with alkali and use reverse transcriptase to duplicate strand

3. Cut "hairpin" with S1 nuclease

cDNA for insulin

4. Add synthetic Hind III linkers by blunt-end ligation

CCAAGCTTGG ——— CCAAGCTTGG
GGTTCGAACC ——— GGTTCGAACC

5. Cut gene and pBR322 plasmid with Hind III to produce "sticky ends"

AGCTTGG ——— CCA
ACC ——— GGTTCGA

pBR322

6. Mix and anneal with T4 ligase

Insulin gene pBR322

Figure 5.14 The first insulin clone, as made by Rutter and Goodman.

the total mRNA was mixed with reverse transcriptase and cDNAs were synthesized. The primer attaches by hydrogen bonding to the mRNA and provides a starting point for reverse transcriptase. To increase the probability that preproinsulin mRNA would be copied, the primer sequence (10 nucleotides long) was made complementary to a known sequence found at the beginning of the insulin mRNA.

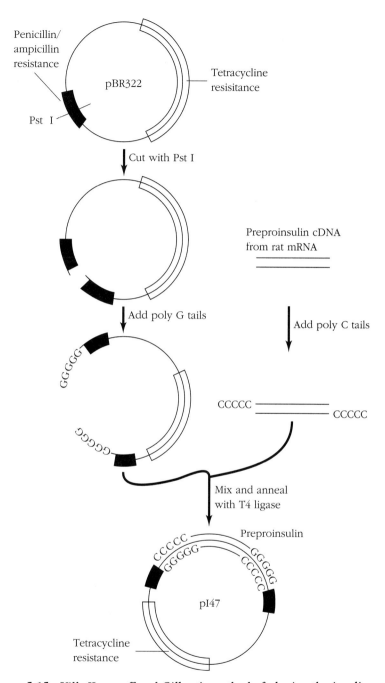

Figure 5.15 Villa-Komaroff and Gilbert's method of cloning the insulin gene.

The vector the Gilbert group decided to use was pBR322, but they added an extra twist to this already versatile plasmid (Figure 5.15). Using the restriction endonuclease Pst I, they cut into the gene for penicillin/ampicillin resistance. Then they added poly G tails to the free ends of the opened pBR322. With the enzyme terminal transferase, poly C tails were added to the preparation of cDNAs. When the preproinsulin cDNA and plasmid were mixed together with T4 DNA ligase, the preproinsulin gene fused with the penicillin/ampicillin-resistance gene. The latter gene codes for an enzyme, penicillinase, that destroys the activities of penicillin and ampicillin. The Gilbert group predicted that a fused hybrid protein, penicillinase-preproinsulin, would be produced by this plasmid. Since penicillinase is normally secreted by bacteria that carry the penicillin-resistance gene, they hoped that preproinsulin or, ideally, proinsulin would be secreted along with it.

After transfecting bacteria with the recombinant plasmids, called pI47, they went through several steps to determine which colony expressed the preproinsulin gene. First they probed the colonies with radioactive cDNA that they assumed to be specific for the insulin sequence. The colonies that tested positive were then screened for insulin synthesis. Fortuitously, Stephanie Broome was in the Gilbert lab. She had just developed the "sandwich" technique, a solid-phase radioimmune assay for the detection of minute quantities of a specific protein (see Figure 5.11). Only one modification to her protocol was necessary: in addition to insulin antibodies, they used radioactively tagged penicillinase antibodies. In this way, they could detect clones that produced the proinsulin-penicillinase hybrid protein. Such colonies would darken the film when either insulin- or penicillinase-radiolabeled antibodies were added.

Furthermore, they demonstrated that the hybrid protein was actually being secreted into the *periplasmic space*; that is, the space between the cell wall and the plasma membrane. To show this effect, they had to subject the cells to osmotic shock with ice-cold water. This treatment released proteins, including the hybrid, from the periplasmic space without lysing and killing the cells. Now they could obtain proinsulin from *E. coli* and use the same cells again to produce more of this valuable protein.

The project of Gilbert and his associates, published in August 1978, opened many avenues for future research. Later studies led to the remarkable conclusion that *E. coli* enzymes specifically removed the polypeptide leader (pre-) sequence from preproinsulin during its secretion. The secreted hybrid contained proinsulin joined to penicillinase. Some signals for protein processing in *E. coli* must therefore be very similar to those in the human preproinsulin molecule. Thus, throughout the course of over 3 billion years of evolution, the amino acid sequences that serve as signals for protein-processing enzymes have apparently changed very little.

Constructing the Human Insulin Gene

The first success in cloning the human insulin gene involved a collaborative effort by research teams from Genentech, Inc., and the City of Hope National Medical Center. Their strategy was highly successful from a marketing standpoint and involved several important procedures that illustrate the remarkable creativity that is prevalent in this field of research. An oversimplified flow diagram of this project is shown in Figure 5.16.

Knowing the genetic code and the sequence of the 50 amino acids present in the secreted form of insulin, a small team under the direction of Keichi Itakura started out by synthesizing the DNA that codes for this hormone. In only three months Itakura's group joined all of the triplet oligonucleotide codons in the correct order. Obviously this procedure circumvented the need to isolate the human mRNA that codes for insulin and resulted in a gene that was unmistakably correct. However, this approach was probably taken because the City of Hope lab was geared up for it, having available all of the necessary trinucleotide codons and considerable experience in synthesizing DNA of a defined sequence.

A more accurate representation of the first four steps in the cloning procedure for the A chain is diagrammed in Figure 5.17. A similar strategy was employed for the B chain. The A chain was designed so that it would have an Eco RI site on one end and a Bam HI site on the other. These enzymes were used to treat pBR322, cutting out a segment that included the start of the tetracycline-resistance gene (Tet^R). The ampicillin-resistance gene (Amp^R) remained intact. Because two different restriction sites were employed, the A chain could be inserted into the plasmid in only one orientation. As we shall soon see, this is important for expression of the gene in the bacterial cell. After the A chain was inserted into the plasmid, *E. coli* cells were transformed with it and checked to make sure that the A chain replicated properly. Using Eco RI, the team cut out and isolated part of the *E. coli lactose* (*lac*) *operon*. This segment included the promoter, the operator, and almost all of the β-galactosidase gene (3000 base pairs). With the A chain already present in a plasmid containing a single Eco RI site, it was an easy matter to cut open this plasmid with Eco RI and insert the section of the lac operon. The resulting plasmid now contained DNA that coded for the A chain under the control of the promoter from the bacterial cell's lac operon.

You may wonder why the researchers took the trouble to insert the lac operon. They did so for several very good reasons, using their ingenuity to make things as simple as possible. First, they could control expression of the A and B chains by growing the bacterial colonies on lactose, because the Z gene of the lac operon codes for β galactosidase, an enzyme that is produced only when the medium contains lactose and little or no glucose. Second, colonies that produce β galactosidase can be detected by growth on X-gal, a compound (5-bromo-4-chloro-3-indolyl-β-D-galactoside) that resembles lactose so closely that β galactosidase will break it down. One product of this hydrolysis reaction is blue.

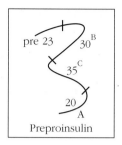

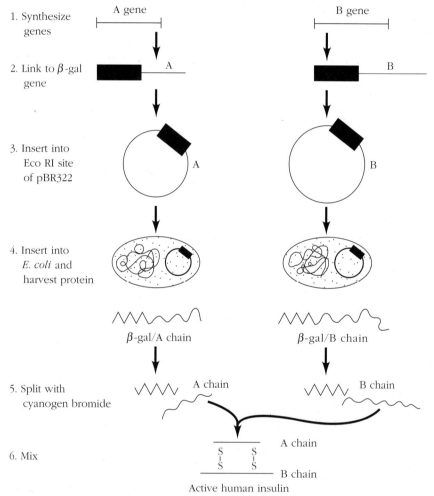

Figure 5.16   Genentech's method of cloning the insulin gene.
The β-gal gene codes for β-galactosidase in *E. coli*. This gene provides an easy
method for the detection of recombinant clones.

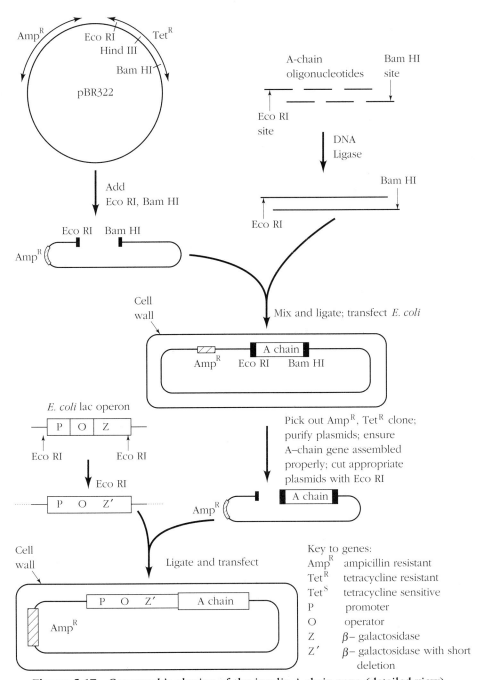

Figure 5.17 Genentech's cloning of the insulin A-chain gene (detailed view).

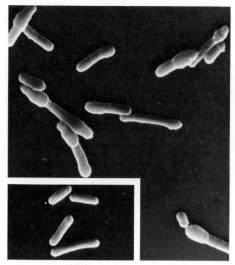

Figure 5.18 Bacteria engineered to produce insulin.
The bulges in the cells are caused by large amounts of insulin. The inset
shows the same strain without the insulin plasmid.

The degree of color varies with the amount of β galactosidase produced.
Combining this indicator with the ampicillin-resistance marker and tetracycline sensitivity creates a very good method for sorting out colonies
that possess the desired plasmid. The third reason for inserting the lac
operon next to both the A and B chains was based on the knowledge that
a foreign protein such as insulin is rapidly degraded by bacteria. The
investigators correctly postulated that a hybrid protein, composed partly
of insulin and partly of the bacterium's own β galactosidase, would help
protect the foreign protein (that is, insulin) from this degradation
process.

  We mustn't overlook one more clever innovation. A synthetic
codon for methionine was introduced between the lac operon and the
insulin A and B chains. This process permitted the release of both the A
and B chains of insulin from the hybrid protein by means of cyanogen
bromide treatment (Figure 5.16, step 5). Cyanogen bromide is commonly used to cleave proteins specifically at methionine residues. Once
the insulin chains were released, some fairly straightforward protein
chemistry was used to get the A and B chains to join by forming the
correct disulfide bridges (step 6).

  On August 21, 1978, the first genetically engineered human insulin
was announced. Figure 5.18 shows bacteria bulging with insulin. Although this procedure was quite clever, it was far from being economi-
cally practical. Nevertheless, it was enough to persuade Eli Lilly to sign a
contract with Genentech. By 1982 the Food and Drug Administration
had approved Humulin, Eli Lilly's genetically engineered insulin. Soon
thereafter, this company had gained control of 90 percent of the American market for insulin.

5.4 Other Applications of Genetic Engineering Technology

As we have seen, there are powerful techniques for cloning a gene and introducing it into a bacterium in a form that will replicate and produce large quantities of its encoded protein product. Let us take a brief look at the way these techniques are being used, with some ingenious variations, in two other areas of research that have profound scientific and commercial potential.

5.4.1 Detecting Human Genetic Diseases Before Birth

With over 100,000 genes present in each of our cells, it is not surprising that every human bears a large number of deleterious mutations. Many of these mutations are serious enough to be lethal were it not for the fact that we are diploid organisms, and in most cases a single "good" copy of a gene codes for the production of sufficient protein for survival and even normality. However, about 3000 hereditary disorders are known to be caused by mutations in single genes, and they occur in approximately 1 percent of newborns.

Unlike abnormalities that arise from improper chromosomal segregation (*nondisjunction*), such as trisomy 21, XXY, and XXX, *point mutations* (mutations in a single base pair) and small deletions within a gene are too small to be detected prenatally or even after birth by karyotypic (chromosomal) analysis. Nearly 100 single-gene disorders can be detected by "conventional" means — that is, by extraction of fetal cells by a procedure such as amniocentesis or *chorionic villi sampling*. The fetal cells are grown in culture until there is a sufficient number to identify an abnormal or deficient gene product (protein) by standard biochemical methods. For the vast majority of single-gene defects, however, it is not yet possible to use this approach because the defective gene product either is not known or cannot be assayed.

Restriction endonucleases and related biotechnology are being used for the detection of genetic defects that cannot be seen by conventional means. The classic genetic strategy, known as *linkage analysis*, has been revitalized by this modern technology. The restriction enzymes have revealed *genetic markers*, DNA fragments of a precise molecular length, which are closely linked to the mutated genes that are responsible for inherited diseases. This technology permits the mutant gene to be localized to a position near the marker and thus enables the molecular geneticist to clone the gene and study its activity. Moreover, the inheritance of the gene can be followed by virtue of this genetic linkage, thereby opening the door to simple tests for detecting heterozygous carriers and future victims of the disease.

This methodology was first used in the prenatal diagnosis of a very rare form of anemia called homozygous $\delta\beta$ *thalassemia*. Thalassemia

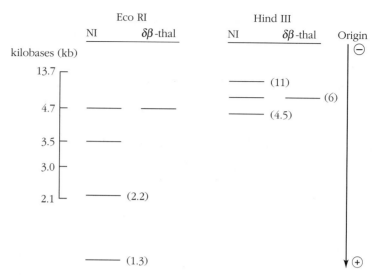

Figure 5.19 Idealized Eco RI and Hind III restriction maps of DNA from normal subjects (NI) and persons homozygous for $\delta\beta$ thalassaemia. A β-globin cDNA sequence was used as the radiolabeled probe.

appears in many forms, all characterized by a reduced rate of synthesis of one or more of the various types of protein chains (*globins*) found in hemoglobin. Several genetic loci determine the synthesis of these globins, and their expression varies during development of the embryo and even after birth. While all hemoglobins contain two α chains, the other two globin chains may be γ, δ, ϵ, or β, depending on age. Thalassemias occur widely throughout the Middle East, the Indian subcontinent, and southeast Asia, and are probably the most common single-gene disorder in the world. They are classified into four genetic types: α, β, $\delta\beta$, and $\gamma\delta\beta$, depending on which globins are affected.

The case of $\delta\beta$ thalassemia is easiest to understand with respect to its detection by restriction endonuclease mapping. As we shall see, this type differs from most in that the lack of globin chain synthesis results from a deletion of the δ- and β-globin structural genes. Restriction enzyme digestion of DNA from normal persons and from persons homozygous for $\delta\beta$ thalassemia yielded DNA fragments that were separated by gel electrophoresis (Figure 5.19). The DNA in the gels was then transferred to nitrocellulose filter paper and subjected to Southern blot analysis, described earlier (see Figure 5.4). The probe in this case was a P^{32}-radiolabeled β-globin cDNA sequence, prepared by reverse transcription of the purified mRNA. Figure 5.19 shows idealized data for digestions by Eco RI and Hind III.

It is evident that with both enzymes only one of the three or four bands seen with the normal subjects was present in the patients with $\delta\beta$ thalassemia. The absence of the other bands provided clear evidence that genes homologous to the probe were deleted in these homozygous patients and readily accounts for the complete deficiency in production of the δ and β globins. In the case of *heterozygotes*, the DNA bands that were totally absent in the homozygotes were half as dark as the other bands in the autoradiogram, indicating that the heterozygotes had one normal and one partially deleted chromosome.

This technique is applicable to point mutations as well as to deletions and was first applied successfully to the prenatal detection of another common blood disorder, sickle-cell anemia. This disease, caused by a point mutation in the gene that codes for the β-globin chain of hemoglobin, is found predominantly in people whose ancestry can be traced to regions that have had severe epidemics of malaria, as we saw in Chapter 1. Briefly, the procedure for detecting the mutation responsible for sickle-cell anemia was as follows (see Figure 5.20). DNA from persons with no known ancestors from these regions was treated with Hpa I restriction endonuclease and the resulting fragments were separated by electrophoresis, then hybridized with a radiolabeled probe containing

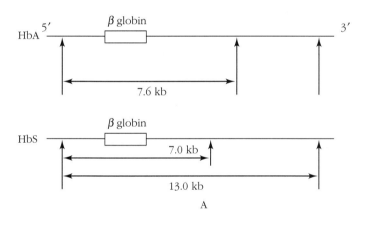

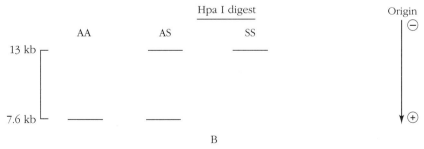

Figure 5.20 Hpa I restriction fragments.
(A) Representation of the three types of restriction fragments observed containing the β and sickle globin genes. (B) Diagrammatic representation of the restriction fragment pattern of the autoradiograph of the Southern blots of the Hpa I digests of DNA from normal persons (AA) and persons with sickle-cell trait (AS) and disease (SS) probed with a radiolabeled β-globin cDNA sequence.

the β-globin sequence. Autoradiography indicated that there was a 7.6-kilobase (kb) fragment, which by DNA sequence analysis was known to contain the entire β-globin structural gene and a portion of the 5′ and 3′ flanking regions. The DNA obtained from most persons of West African origin did not contain this 7.6-kb restriction fragment, but contained instead either a 7.0-kb or a 13.0-kb fragment. These DNA fragments of different sizes found in different persons, all of which contain sequences that are homologous to a specific probe (β globin, in this case), are referred to as *restriction fragment length polymorphisms*, or RFLPs. In the case of the sickle-cell trait, the RFLPs result from a single base-pair change (point mutation) near the β-globin gene. One such mutation creates a new Hpa I site within the 7.6-kb fragment, causing it to become shortened to 7.0 kb. Another abolishes the Hpa I site at the 3′ end, causing the globin-containing fragment to increase in size to 13.0 kb.

It turns out that 80 percent of West Africans who suffer from sickle-cell anemia also carry one of the two unusual Hpa I RFLPs. This finding (that is, linkage between the rare RFLPs and the disease) implies that the mutation that causes this disease arose in the DNA of a person in which the Hpa I RFLP was already present. The close proximity of these two mutations, one causing sickle-cell anemia and one causing the unusual RFLP, prevents them from becoming separated from each other through *meiotic recombination*. Thus the Hpa I RFLP serves as a fairly accurate genetic marker for persons who are either homozygous or heterozygous for the sickle-cell mutation.

Theoretically this approach can be taken for prenatal diagnosis of any inherited disease. However, it is necessary to find an RFLP marker that is linked closely enough to the disease-causing mutation in question, and this is no simple matter. Note that only 80 percent of West Africans who carry the sickle-cell trait also possess one of the two unusual RFLPs. The closer the linkage between the restriction mutation and the mutation that causes the disease, the better this method can predict the disease. "Reasonably tight linkage," point out R. White and J.M. Lalouel (1988, p. 40), "can make the marker useful for diagnosing the disease in members of an afflicted family; very tight linkage can open the way to identification and cloning of the defective gene." A short list of genetic diseases for which the chromosomal location of the defective gene has been determined by restriction enzyme analysis is shown in Table 5.1. The sequencing of the entire human genome is currently a major worldwide project. If it succeeds, we can hope to detect all heritable diseases before an afflicted child is born.

5.4.2 Introducing Human Genes into Mice

Whereas bacteria are wonderful factories for the economical production of large quantities of a particular gene product, prokaryotic

Table 5.1. Nine genetic disorders from which chromosomal location of the defective gene has been determined by linkage studies.

Disease	Chromosome carrying defective gene and linked marker	Year determined
Huntington's disease	4	1983
Duchenne muscular dystrophy	X	1983
Polycystic kidney disease	16	1985
Cystic fibrosis	7	1985
Chronic granulatomatous disease	X	1985
Peripheral neurofibromatosis	17	1987
Central neurofibromatosis	22	1987
Familial polyposis coli	5	1987
Multiple endocrine neoplasia IIIa	10	1987

organisms are of limited use. For example, in studies concerning the developmental regulation of gene expression, so that a gene is turned on or off or fine-tuned as the embryo develops into a newborn, then into a young animal, and finally into an adult, there are obvious advantages in introducing the cloned gene into the germ line of a mouse rather than into that of a human. Such a procedure was first carried out successfully in 1980 with a gene from herpes simplex virus and also with rabbit and human β-globin genes. Mice bearing heritably transmissible foreign genes are referred to as *transgenic mice*. Before considering some of the advantageous applications of this area of research, let us first briefly see how transgenic mice are produced (Figure 5.21).

Fertilized eggs are usually collected from the oviducts of female mice that have been hormonally primed for ovulation of large numbers of eggs. Highly purified cloned DNA is injected into the undivided nuclei of these eggs with an ultrafine needle (about 1 μm at the tip). Considerable care is needed to avoid lysing the egg cell and to insert the DNA in the right place. After a brief incubation in culture, the eggs are transferred directly into the oviduct of a pseudopregnant mouse. Before receiving the eggs, the foster mother was prepared for pregnancy by mating with sterile (vasectomized) males. The newly introduced gene integrates either as a single copy or as multiple copies randomly in the offspring's chromosomes. After six to eight weeks, the transgenic mice become sexually mature and ready to pass their newly acquired gene(s) on to their offspring.

One of the earliest of these remarkable projects resulted in the creation of a transgenic mouse carrying a rat gene that codes for the production of a growth hormone. As is evident in Figure 5.22, the newly acquired gene was efficiently expressed. At 10 weeks of age, the transgenic mouse (on the left) weighed 44 g while his normal sibling

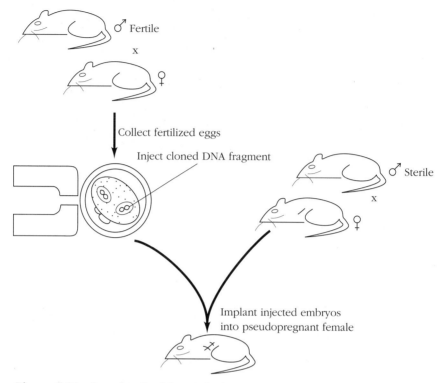

Figure 5.21 Steps involved in producing a transgenic mouse. Fertilized eggs are collected, injected with cloned DNA, and transferred to a pseudopregnant foster mother.

weighed in at 29 g. On the average, mice that expressed the rat growth hormone gene grew two to three times faster than controls, and reached up to twice the size of a normal mouse.

The applications of transgenic mice are of considerable value. Consider, for example, the field of biomedical research. Imagine being able to use mice instead of humans to study the roles of genes that cause human diseases. Genes from viruses that cause such dreaded diseases as AIDS and rabies can be introduced into mice, and their role in the development of the disease can be investigated. The expression of cellular oncogenes is known to be involved in cancer, but the mechanism is very complex. Different tissue types are affected differently by the presence of a given oncogene, and whether the oncogene produces a tumor can depend on whether it is expressed before or after the cells have differentiated. Thus the susceptibility of a given cell to becoming tumorigenic depends to some extent on the developmental stage of that cell. Genes that increase the probability of a particular type of cancer,

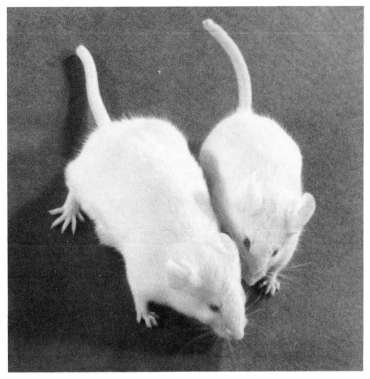

Figure 5.22 Changes in mouse physiology by introduction of DNA into the germ line.
The mouse on the left contains a new gene composed of the mouse metallothionein promoter and regulatory region fused to the rat growth hormone structural gene. He weighs 44 g at 10 weeks of age and is shown next to a sibling weighing 29 g.

such as retinoblastoma, are also being introduced and studied in mice. The introduction of oncogenes into mice is enabling scientists to achieve considerable progress in their search for the molecular basis of cancer.

Advances in the detection and treatment of diseases can also be made more readily with mice than with humans. In many cases, genetic analyses permit the identification of a gene that is responsible for a certain disease long before the protein product of that gene has been determined. When such genes are used to create transgenic mice, the animals can be used to develop diagnostic procedures for the early detection of the disease and for chemotherapeutic treatments and other procedures that could alleviate symptoms and eventually lead to a cure for the disease in humans.

Discussion Questions

1. Briefly describe the Southern blot method used to detect specific DNA sequences in a polyacrylamide gel after electrophoresis.

2. What five important properties make a plasmid suitable as a vector for a newly inserted gene?

3. Describe the use of the β-galactosidase gene for the detection of colonies containing a desired recombinant gene inserted into a plasmid.

4. How does the presence of two antibiotic-resistance genes (such as Tet^R and Kana^R) in a single plasmid enable the researcher to identify easily a desired gene inserted in that plasmid?

5. a. What is a cDNA library?
   b. How is a cDNA library obtained experimentally?

6. a. What is the shotgun method for obtaining clones of desired genes?
   b. How do clones obtained from eukaryotes by the shotgun procedure differ from those obtained by preparation of a cDNA library?

7. How are hot probes used to identify bacterial colonies containing a newly inserted desired gene?

8. How can antibodies to insulin be used to detect a colony that is producing insulin?

9. a. What are restriction fragment length polymorphisms (RFLPs)?
   b. Explain why people with a particular genetic disease, such as sickle-cell anemia, might exhibit unusual RFLPs.

10. What is a transgenic mouse and how may it be useful to humankind?

References

Berg, P. (1981). Dissections and reconstructions of genes and chromosomes. *Science, 213*:296.

Camper, S.A. (1987). Research applications of transgenic mice. *BioTechniques, 5*:638.

Cherfas, J. (1982). *Man-Made Life.* New York: Pantheon Books.

Cohen, S. N. (1975). The manipulation of genes. *Scientific American, 233*:24.

Cohen, S. N., A. C. Y. Chang, H. W. Boyer, and R. B. Helling (1973). Construction of biologically functional bacterial plasmids *in vitro. Proceedings of the National Academy of Science USA, 70*:3240.

Gilbert, W., and L. Villa-Komaroff (1980). Useful proteins from recombinant bacteria. *Scientific American, 242*:74.

Goeddel, D. V., D. G. Kleid, F. Bolivar, H. L. Heyneker, D. G. Yansura, R. Crea, T. Hirose, A. Kraszewski, K. Itakura, and A. D. Riggs (1979). Expression in *Escherichia coli* of chemically synthesized genes for human insulin. *Proceedings of the National Academy of Sciences USA, 76*:106.

Grunstein, M., and D. S. Hogness (1975). Colony hybridization: a method for the isolation of cloned DNAs that contain a specific gene. *Proceedings of the National Academy of Sciences USA, 72*:3961.

Gueriguian, J. L. (ed.) (1981). *Insulins, Growth Hormone, and Recombinant DNA Technology.* New York: Raven Press.

Itakura, K. (1980). Synthesis of genes. *Trends in Biochemical Science, 5*:114.

Itakura, K., and A. D. Riggs (1980). Chemical DNA synthesis and recombinant DNA studies. *Science, 209*:1401.

Lederberg, J., and E. M. Lederberg (1952). Replica plating and indirect selection of bacterial mutants. *Journal of Bacteriology, 63*:399.

Macleod, A., and K. Sikora (eds.) (1984). *Molecular Biology and Human Disease.* Oxford, England: Blackwell Scientific Publications.

Morrow, J. F., S. N. Cohen, A. C. Y. Chang, H. W. Boyer, H. M. Goodman, and R. B. Helling (1974). Replication and transcription of eukaryotic DNA in *Escherichia coli. Proceedings of the National Academy of Sciences USA, 71*:1743.

Mueller, C., A. Graessmann, and M. Graessman (1980). Microinjection: turning living cells into test tubes. *Trends in Biochemical Science, 5*:60.

Nathans, D., and H. O. Smith (1975). Restriction endonucleases in the analysis and restructuring of DNA molecules. *Annual Review of Biochemistry, 44*:273.

Novick, R. P. (1980). Plasmids. *Scientific American, 243*:102.

Okayama, H., and P. Berg (1982). High-efficiency cloning of full-length cDNA. *Molecular and Cell Biology, 2*:161.

Prentis, S. (1984). *Biotechnology: A New Industrial Revolution.* New York: George Braziller.

Southern, E. M. (1975). Detection of specific sequences among DNA fragments separated by gel electrophoresis. *Journal of Molecular Biology, 98*:503.

Temin, H., and D. Baltimore (1972). RNA directed DNA synthesis and RNA tumor viruses. *Advances in Virus Research, 17*:129.

Ullrich, A., J. Shine, J. Chirgwin, R. Pictet, E. Tischer, W. J. Rutter, and H. M. Goodman (1977). Rat insulin genes: construction of plasmids containing the coding sequences. *Science, 196*:1313.

Villa-Komaroff, L., A. Efstratiadis, S. Broome, P. Lomedico, R. Tizard, S. P. Naber, W. L. Chick, and W. Gilbert (1978). A bacterial clone synthesizing proinsulin. *Proceedings of the National Academy of Sciences USA, 75*:3727.

White, R., and J.-M. Lalouel (1988). Chromosome mapping with DNA markers. *Scientific American, 258*:40.

Williams, D. C., R. M. Van Frank, W. L. Muth, and J. P. Burnett (1982). Cytoplasmic inclusion bodies in *Escherichia coli* producing biosynthetic human insulin proteins. *Science, 215*:687.

1

Techniques and Procedures

Listed in order of appearance in text.

Chapter 1. The Power of Microscopy
Various types of microscopy (light, phase contrast, interference, scanning electron, transmission electron)
Fluorescent antibody staining of cellular proteins/phase contrast or fluorescence microscopy
Freeze-fracturing, freeze-etching, metal shadowing
Cell fusion
Electrical field applied to organelles
Receptor localization by fluorescent antibody tagging
Length analysis of DNA by electron microscopy
Autoradiography
Chromosome banding
Karyotypic analysis (karyology)

Chapter 2. Purification and Properties of Protein Kinase
Differential centrifugation
Column chromatography (ion exchange, gel filtration, affinity, and high-performance liquid chromatography)
Enzyme assays
Acid precipitation of macromolecules; filtration
Analysis of radioactivity by liquid scintillation counting
Salting-in and salting-out proteins
Simultaneous multiple-column elution of proteins by means of a single gradient forming apparatus
Polyacrylamide gel electrophoresis (PAGE)
Immunoprecipitation/PAGE/autoradiography
Electrofocusing (isoelectric focusing)
Use of temperature-sensitive (ts) mutant to identify gene product

Chapter 3. The Discovery and Properties of Messenger RNA
Electron microscopy
Equilibrium density gradient centrifugation (in CsCl)
Ultraviolet light (UV) absorption analysis of optical density (OD) with a
  spectrophotometer
Pulse-chase radiolabeling and density labeling of macromolecules
DNA-RNA hybrid formation
Double labeling: discriminating two radioisotopes in a single sample by liquid
  scintillation counting
Rate-zonal centrifugation (sucrose gradients)
Use of inhibitors in kinetic analysis of macromolecular synthesis and deg-
  radation
Macromolecular size distribution analysis of RNA by sedimentation rate
Measurement of DNA lengths involved in RNA synthesis by electron microscopy
  of DNA-RNA hybrids (R-loop analysis)
Visualization of RNA synthesis by electron microscopy

Chapter 4. Cell Differentiation, Aging, and Cancer: Cell Culture Studies
Culturing of cells and tissues
Cytochemical detection of enzyme
Scanning electron microscopy for identification of cell type
Autoradiography of metaphase chromosomes
Assays for transformed cells (focus; colony formation in soft agar)
Assay for tumorigenic cells by growth in athymic ("nude") mice
Karyotypic analysis; G-banding
Cell fusion
Creation of hybridoma cell lines/monoclonal antibodies

Chapter 5. Genetic Engineering: Techniques and Applications
Restriction endonuclease digestion of DNA
Polyacrylamide gel electrophoresis (PAGE) of DNA
Southern blot analysis of homologous DNA sequences
Formation of recombinant DNA (synthetic linkers, ligation, etc.)
Creating a cDNA library (cDNA cloning)
Genomic DNA cloning ("shotgun" method)
Gene synthesis from mRNA
Replica plating of bacterial colonies
Clone detection by cDNA or RNA hybridization
Clone detection by use of labeled antibodies directed against gene product/
  radioimmune assay
Clone selection by use of plasmid containing two drug-resistance genes
Regulate gene expression by fusion to *E. coli's* lactose operon
Introduction of a synthetic codon to facilitate cleavage of desired gene product
  from protein to which it is fused
Detection of genetic diseases before birth by restriction fragment length poly-
  morphisms (RFLP)
Production of transgenic mice

APPENDIX

2

Abbreviations and Acronyms

Å	angstrom
Ab	antibodies, antiserum
Ag	antigen
AIDS	acquired immunodeficiency syndrome
AMP	adenosine monophosphate
Amp^R	ampicillin resistance
ARC	AIDS-related complex
ASV	avian sarcoma virus
ATP	adenosine triphosphate
bp	base pairs
BSA	bovine serum albumin
cAMP	cyclic AMP
cDNA	complementary DNA
CMP	cytidine monophosphate
CP	chloramphenicol
CPK	creatine phosphokinase
CPM	counts per minute
d	daltons
DEAE	diethylaminoethyl
DNA	deoxyribonucleic acid
DNP	deoxyribonucleoprotein
EBV	Epstein-Barr virus
EM	electron microscope
ESR	electron spin resonance
FCS	fetal calf serum
FITC	fluorescein isothiocyanate
GMP	guanosine monophosphate
GPh	glycogen phosphorylase
HbS	hemoglobin S
HESM	human embryonic skin and muscle
HIV	human immunodeficiency virus
hnRNA	heterogeneous nuclear RNA

HPLC	high-performance liquid chromatography
Kana^R	kanamycin resistance
kb	kilobase
lact	lactalbumin
MK	myokinase
M_r	molecular weight
mRNA	messenger RNA
nd	nondefective
NGF	nerve growth fiber
nm	nanometer
NMR	nuclear magnetic resonance
NPF	neurite promoting factors
NTF	neuronotrophic factors
OD$_{254}$	optical density at 254 nm
oligo dT	oligodeoxythymidylic acid
PAGE	polyacrylamide gel electrophoresis
PHA	phytohemagglutinin
pI	isoelectric point
poly A	polyadenylic acid
poly T	polythymidylic acid
R	rhodamine
RBCs	red blood cells
rDNA	genes encoding ribosomal RNA
RFLPs	restriction fragment length polymorphisms
RNA	ribonucleic acid
RP	resolving power
rpm	revolutions per minute
rRNA	ribosomal RNA
RSV	Rous sarcoma virus
SDS	sodium dodecyl sulfate
SR-ASV	Schmidt-Ruppin strain of ASV
SV40	simian virus 40
TCA	trichloroacetic acid
TdR	thymidine
Tet^R	tetracycline resistance
TMR	tetramethylrhodamine
ts	temperature sensitive
U	enzyme unit
UMP	uridine monophosphate
UV	ultraviolet

Glossary

active site The peptide region of an enzyme that is involved in the conversion of a substrate to a product.

agarose A naturally occurring polysaccharide (sugar polymer) commonly prepared commercially for use in column chromatography and gel electrophoresis.

alleles Different forms of a particular gene, generally resulting from mutations.

allosteric site The peptide region of an enzyme that interacts with a regulatory molecule, making the enzyme either more or less active; a region removed from the active site, generally on a different polypeptide chain of the enzyme.

amino terminal end Polypeptide chains, if not circular, possess a free amino group at one end (the amino terminus) and a free carboxyl group at the other end; the synthesis of each protein begins at the amino terminal end.

amniocentesis A technique for determining genetic abnormalities in a fetus or embryo; involves removal of fetal cells that have been sloughed off into the amniotic fluid in which the fetus is developing.

ampholyte A mixture of small charged organic molecules that can form a linear pH gradient when they are present in a gel that is subjected to an electrical current.

aneuploid Without the normal number of chromosomes.

annealing (nucleic acid hybridization) The process by which two single-stranded polynucleotides form a double-stranded molecule, with hydrogen bonding between the complementary bases of the two strands. Annealing can take place between complementary strands of either DNA or RNA to produce double-stranded DNA molecules, double-stranded RNA molecules, or RNA-DNA hybrid molecules.

antibiotics Substances, such as penicillin and erythromycin, that have anticellular properties, through inhibition of reproduction and/or metabolic functions.

antibody A protein produced by B lymphocytes that binds specifically to a foreign protein (antigen) as a first step in eliminating an antigen from the body.

antigen A chemical (generally foreign to the body) that elicits an immune response; that is, stimulates the immune system to produce antibodies that specifically bind to the antigen.

antiserum Serum containing a specific type of antibodies.

axon An outgrowth of a nerve cell that conducts signals away from the cell body.

bacterial colony A cluster of bacteria that originates from the division of a single cell; a bacterial clone.

bacteriophage (phage) A virus whose host is a bacterial cell.

Barr body A highly condensed, inactivated X chromosome observed in female mammalian cell nuclei.

B cells Cells in the pancreatic islets of Langerhans, which secrete insulin; also called beta (β) cells.

B lymphocyte A white blood cell that originates and matures in the bone marrow and is responsible for humoral immunity; that is, immunity resulting from antibodies circulating in the bloodstream and lymphatic system. *See* **lymphocyte**.

budding The process by which certain membrane-enclosed viruses are released from the plasma membrane of eukaryotic cells.

capsid The protein coat of a virus.

carcinogen A chemical that can cause cancer.

casein A protein abundant in milk and commonly used as a substrate for protein kinases.

cDNA library A collection of DNA sequences that are complementary to all or most of the mRNA from a cell.

cell line A specific type of cell that has become immortal and can be grown in cell culture indefinitely; most cell lines can be stored frozen.

chloramphenicol (CP) An antibiotic that inhibits protein synthesis in susceptible bacteria.

chorioallantoic membrane A membranous cell layer surrounding a chick embryo; useful in the assay of certain types of virus, such as the influenza virus.

chorionic villi sampling A procedure for diagnosing the genetic abnormalities in an embryo or fetus; a portion of the placenta is removed and analyzed.

chromatid One of the paired, virtually identical strands of a chromosome after it has replicated its DNA.

chromatin Material composed of partially condensed DNA and basic proteins (histones) found in all eukaryotic cells; can condense further into chromosomes during mitosis and meiosis.

chromosome The most highly condensed form of chromatin; most visible and present in the cell nucleus during mitosis and meiosis.

cilium (pl. cilia) A specialized cellular appendage composed of proteins and used for locomotion.

citric acid cycle *See* **Krebs cycle**.

clone A lineage of genetically identical DNA molecules, cells, or organisms.

codons A group of three adjacent nucleotides in mRNA and DNA which code either for a specific amino acid or for polypeptide chain termination during protein synthesis.

competitive inhibitor A molecule that competes with a substrate for the active site on an enzyme.

complementary Pertaining to genetic material that has nitrogenous bases arranged in a sequence that "matches" or pairs with the material it complements; that is, each adenine paired with thymine or uracil, each guanine paired with cytosine, and vice versa. Pairing occurs by hydrogen bonding between the bases.

conditional mutant A mutant cell or virus that exhibits a particular phenotype under one set of conditions (such as low temperatures) but not under another (such as high temperatures).

crisis A genetic change that occurs in a large number of cells after repeated passaging; results in the loss of proliferative capacity and a change in the number of chromosomes.

cytosol The fluid portion of the cytoplasm.

denature To disrupt the structure of proteins and nucleic acids by heat, detergents, pH, etc., which break ionic and hydrogen bonds.

dendrite The branched receiving process of a nerve cell.

differentiation The process of becoming more highly specialized; for example, during embryonic development, as cells become organized into tissues and organs.

diploid cell A cell that has two sets of chromosomes, one set inherited from each parent.

dorsal The top or back half of a bilaterally symmetrical animal.

eluant A solution used to wash off a substance that is bound to a resin or other type of solid matrix; used in column chromatography.

endocytosis The process by which macromolecules and particles are engulfed and ingested by localized regions of the plasma membrane, forming intracellular vesicles; some endocytosis is mediated by receptor proteins.

endogenous Already present within a particular sample.

endoplasmic reticulum A cytoplasmic membranous network involved in protein synthesis, membrane production, the discharge of transport vesicles (membrane-coated packages of substances), and many other functions.

enzyme A biological catalyst, generally a protein; speeds up the rate of a chemical reaction without being altered at the end of the reaction (Note: some RNA molecules have recently been shown to possess enzymatic activity involved in the formation of mRNA from hnRNA.)

epithelial cells Cobblestone-shaped cells that form sheets that line organs and body cavities; cells derived from the embryonic ectodermal or endodermal cell layers

eukaryotic cell A type of cell that has a membrane-enclosed (true) nucleus and other membrane-enclosed organelles.

exons The DNA of a eukaryotic transcription unit whose transcript becomes a part of the mRNA produced by elimination of the introns; contain the coding information for proteins.

fibroblasts Spindle-shaped cells derived from the embryonic mesoderm

fixing As applied to cells, a chemical treatment that dehydrates the cells and denatures macromolecules, causing them to become fixed in position.

fluor An organic solute that gives off fluorescent light when struck by a radioactive emission such as an electron (β particle).

focus A clone of cells that originates from the transformation of a single cell; a cluster of rounded refractile cells that pile up on each other as a result of transformation.

gamete Haploid egg or sperm cells produced by meiosis in animals.

ganglia Aggregations of nerve cell bodies in a centralized nervous system.

gene amplification The repeated replication of certain genes in a chromosome before mitosis.

gene expression The degree to which a particular gene is expressed in a cell in the form of RNA.

gene family A group of genes that are sufficiently related in sequence that they appear to have originated from a common ancestral gene.

generation time The time required for cells in culture to double in number when growing in log phase.

genetic marker An allele whose inheritance is under observation.

genome The entire set of genes (genetic content) in a particular cell or virus.

genotype The genetic constitution of an organism.

glial cells (glia) Cells in the nervous system that provide metabolic support, insulation, and protection for neurons.

glycolytic pathway (glycolysis) A metabolic pathway whereby glucose is converted to pyruvate and ATP (adenosine triphosphate) is formed.

Golgi complex (apparatus) An organelle that looks like stacks of membranes; chemically modifies proteins and helps direct them to their ultimate destination.

haploid A cell, such as a gamete, that possesses only one set of chromosomes (half that of diploid).

hemoglobin A protein produced by red blood cells and composed in adult humans of four protein chains, two called alpha (α) and two called beta (β), which are associated with four heme groups, each bound to an iron atom; used for the transport of oxygen to all the tissues in the body.

hemoglobin S (Hb S) Hemoglobin with the mutated form of the β protein that is found in persons with sickle-cell anemia and in heterozygous carriers of the sickle-cell trait.

heterogeneous nuclear RNA (hnRNA) The primary RNA transcripts found in the nucleus of a eukaryotic cell. The connecting of exons and elimination of introns in these transcripts leads to the formation of mature messenger RNAs.

heterokaryon A single cell with more than one nucleus.

heterozygous Having two different alleles for a given trait.

histone A basic protein associated with DNA in eukaryotic chromatin and chromosomes. *See* **nucleosome**.

homologous With regard to DNA and RNA, synonymous with *complementary*.

homologous chromosomes Chromosome pairs that possess genes for the same traits at corresponding positions (loci). A human cell from a female possesses 23 pairs of homologous chromosomes; each pair is derived from the mother and father. The X and Y chromosomes of a male do not constitute homologous chromosomes.

homozygous Having two identical alleles for a given trait.

hybridization *See* **annealing**

hybridoma A hybrid cell created in the laboratory by the fusion of a tumorigenic lymphocyte (myeloma) with a normal antibody-producing lymphocyte; produces a single (*monoclonal*) type of antibody in cell culture.

hydrophilic Having an affinity for water; soluble in water.

hydrophobic Having an aversion to water; insoluble in water.

hypodiploid Having less than the normal diploid number of chromosomes.

hypotonic shock Disruption of cells by a hypotonic solution; the sudden change in solute concentration causes a rapid influx of water, which breaks the plasma membrane.

hypotonic solution A solution that is lower in solute concentration than that present inside a particular cell.

immunoprecipitation The formation of an aggregate of antigens and antibodies that is large enough to come out of solution, generally upon low-speed centrifugation.

interphase The interval between successive mitoses.

intervening sequences (introns) Noncoding nucleotide sequences in eukaryotic DNA whose RNA transcripts are removed from the primary transcription unit in the process of mRNA formation.

karyology *See* **karyotypic analysis**.

karyotype A photographic display of an entire set of chromosomes from an individual; used to assess abnormalities in the chromosomes, such as number, deletions, translocations.

karyotypic analysis Cytogenetic analysis of the chromosome complement of a cell or organism, including the number, size, and configuration of the chromosomes.

Krebs cycle (citric acid cycle) A metabolic pathway whereby pyruvate is oxidized to carbon dioxide; occurs in mitochondria and when coupled with the electron transport system results in the major production of ATP (adenosine triphosphate) and the reduction of oxygen to water.

lactose (lac) operon Regulates the synthesis of three enzymes, including β galactosidase, required for the use of lactose by certain bacteria (such as *E. coli*).

leukocyte The general name for all white blood cells.

lymphocyte A type of white blood

cell (such as B cell or T cell) that is a major component of the immune defense system; lymphocytes with membrane receptor proteins called CD4 or T4 are susceptible to infection by HIV (human immunodeficiency virus).

lymphoma   A cancer of the lymph nodes, caused by tumorigenic lymphocytes.

lysis   The mechanical or chemical rupture of a cell.

macrophage   A large blood cell that engulfs and degrades other cells; a member of the immune defense system.

medulla   Inner portion of an organ or part.

meiotic recombination   The creation of a new association of DNA molecules (chromosomes) or parts of DNA molecules during the process of meiosis.

mesenchyme   Embryonic connective tissue.

mesoderm   Middle germ layer of the embryo.

messenger RNA (mRNA)   A special type of RNA molecule that brings genetic information from genes to nonspecialized ribosomes; contains the sequence of bases that code for a protein.

metaphase   A stage of mitosis and meiosis at which the chromosomes are fully condensed and lined up with their centromeres along a metaphase plate.

metastasis   The spread of cancer cells to sites distant from the primary tumor, usually via the blood and lymphatic systems.

microfilament   A cytoplasmic fiber 7 nm in diameter constructed principally of the protein actin; involved in intracellular movement and the maintenance of cell shape.

mitochondria   Organelles in eukaryotic cells that are used primarily for cellular respiration.

mitogen   A chemical that stimulates mitosis.

mitosis   A process of cell division in eukaryotes by which the chromosomes duplicate and are equally allocated to each of the two daughter cells.

monoclonal antibodies (monoclonals)   Antibodies that are produced by hybridomas and are specific for a single antigenic site or determinant (epitope).

morphology   Shape; appearance.

myelin sheath   An insulating coat of cell membrane from Schwann cells; insulates axons and helps them propagate nerve impulses.

myoblast   A differentiated cell that is a precursor to a fully differentiated muscle cell.

myosin   A protein that together with actin forms certain types of filaments involved in cellular contraction.

myotube   A multinucleated muscle cell that arises from the fusion of many myoblasts; a terminally differentiated muscle cell.

neurite formation   The sending out of axons and dendrites from neuronal cells; stimulated by nerve-growth factor (NGF), which is actually a neurite-promoting factor.

neurite-promoting factor (NPF)   A substance required for the outgrowth of neuronal processes (axons and dendrites) in culture, and possibly in the animal (*in vivo*).

neuron   A terminally differentiated cell in the nervous system that conducts electrical signals.

neuronotrophic factor (NTF) A substance required for neuron survival in culture, and possibly in the animal (*in vivo*).

nondisjunction Abnormal segregation of chromosomes or chromatids during mitosis or meiosis.

nucleic acid hybridization *See* **annealing.**

nucleolus A specialized region of the chromatin that is used for the synthesis of ribosomal RNA.

nucleosome A particle that can be seen by electron microscopy of lower condensed forms of chromatin; composed of four types of histones around which the DNA wraps.

nude mouse A hairless developmental mutant that is athymic and thus has a very defective immune system.

oncogene A gene whose expression can lead to the conversion of a normal cell into a cancer cell.

oocyte A female gamete in an intermediate stage of maturation (not yet a fully differentiated haploid egg).

operator A nucleotide sequence that is recognized and bound by a repressor protein, which in turn inhibits transcription of the operon.

operon A set of genes consisting of two or more adjacent structural genes, transcribed into a single mRNA, and the adjacent transcriptional control sites (promoter and operator); this organization places the expression of the structural genes in an operon under the coordinate control of a single promoter and operator.

optical density (OD) A measure of the amount of light absorbed by a substance, usually with a subscript indicating the particular frequency of light at which the absorbance occurs.

organelle A membrane-enclosed subcellular structure with one or more specialized functions in eukaryotic cells.

osmotic shock Disruption of cells caused by the movement of water across a selectively permeable barrier such as a plasma membrane or virus protein coat.

palindrome A sequence of symbols that reads identically in both directions: in a palindromic sequence of DNA, the reading of one strand from left to right results in the same sequence found in the other strand when read from right to left.

pandemic A worldwide epidemic of an infectious disease.

passaging Transfer of cells by dilution, generally after they become too crowded in culture or after a fixed number of days of growth.

pellet The solid or semisolid material that forms at the bottom of a tube after centrifugation of a suspension or solution.

periplasmic space The space between the cell wall and the plasma membrane of a cell.

peritoneum lining of the body cavity and covering of the organs.

phage *See* **bacteriophage**.

phenotype The expressed genetic traits of an organism.

phosphodiester backbone The repeating units of phosphate and sugar (ribose in RNA, deoxyribose in DNA) in nucleic acids.

plasma membrane A lipid bilayer surrounding the cytoplasm of all cells; acts as a selective barrier to the passage of ions and molecules.

plasmid   A usually circular genetic element (DNA molecule) harbored within a host cell, which replicates independently of the host DNA.

point mutation   A change at a single nucleotide or base pair within a gene.

polyacrylamide   A relatively stable synthetic polymer commonly prepared commercially for use in gel electrophoresis.

polyribosome (polysome)   A complex consisting of two or more ribosomes joined by their association with a single mRNA molecule.

polytene chromosome   A many-stranded chromosome resulting from repeated replication of the DNA without separation of sister chromatids by mitosis or meiosis.

primary culture   The first transplant of cells from an animal or plant into a growth medium.

probe   A selected sequence of nucleic acids, usually radioactively labeled, that can be used to detect a complementary (homologous) sequence of nucleic acids in a given sample of DNA.

prokaryotic cell   A type of cell lacking a membrane-enclosed (true) nucleus; bacterium.

promoter   A nucleotide sequence in DNA at or near the beginning of a transcription unit which is recognized by RNA polymerase as a binding site.

protein kinases   A class of enzymes that catalyze the transfer of a phosphate from ATP to an amino acid in a protein.

protein serine kinase   A protein kinase that attaches phosphate

from ATP to the hydroxyl group of serine in a protein.

protein tyrosine kinase   A protein kinase that attaches phosphate from ATP to the phenolic group of tyrosine in a protein.

proteolytic enzyme   An enzyme that degrades (hydrolyzes) a protein to its constituent amino acids or to small peptides.

proto-oncogenes   Cellular genes that can be converted into oncogenes via some type of genetic alteration caused by chemicals, radiation, or viruses.

radiolabel   A radioactive chemical; to make radioactive.

receptor protein   A protein found on the surface of a cell and having an affinity for a specific molecule (e.g., insulin receptor).

recessive trait   A phenotypic property that can be completely masked by the expression of a dominant gene in a diploid organism or cell.

recombinant DNA   A DNA molecule composed of DNA from at least two sources.

renaturation   (1) The process by which complementary strands of heat-denatured (single-stranded) DNA can reconstitute the native double-helical DNA structure. (2) The reversion of an enzyme to its native structure.

resolving power (RP)   The minimum distance that two points can be separated and still be distinguished as two different points; related to the power of a lens.

restriction endonuclease   An endonuclease (an enzyme that hydrolyzes internal phosphodiester bonds in a polynucleotide) that recognizes *specific* nucleotide sequences in DNA and then makes a double-strand cleavage of the DNA molecule.

reticulocyte A precursor to a mature red blood cell, with a nucleus still present.

retrovirus A family of RNA-containing viruses (such as avian sarcoma viruses and human immunodeficiency viruses) that contain the enzyme reverse transcriptase within the virion.

reverse transcriptase An enzyme found in retroviruses that copies the viral RNA into a double-stranded complementary DNA (cDNA) in infected cells; this enzymatic activity can be used to prepare cDNA from mRNA in vitro.

revertant A cell that has undergone a mutation that restores the original phenotype.

ribosomal DNA (rDNA) The DNA that is copied (transcribed) into rRNA.

ribosomal RNA (rRNA) The RNA found as a structural component of ribosomes.

ribosome A subcellular particle consisting of RNA and many proteins that functions as the site of protein synthesis.

R-loops The loops created in a DNA-RNA hybrid when there are regions of nonhomology between the two strands.

RNA polymerase An enzyme that assembles a number of subunits of RNA (nucleotides) into a larger unit called an RNA polymer.

sarcoma A malignant tumor that forms in tissues that originate in mesoderm or mesenchyme.

Schwann cells Specialized glial cells that produce a myelin sheath enclosing the axons of many neurons.

scintillation fluid A solvent (such as toluene) used for the detection of radioactive emissions that have been converted to fluorescent light by striking dissolved fluors.

secretory vesicles Cellular "packages" that store substances to be sent outside the cell at a later time. *See* **vesicle**.

senescence As applied to cells, the loss of capacity to divide after repeated passaging; caused by some type of genetic crisis involving a change in chromosome number.

serum The fluid derived from blood after removal of the blood cells.

sickle-cell anemia A disease that causes a deficiency of red blood cells; a result of point mutations in both genes that code for the β protein chain of hemoglobin.

sodium dodecyl sulfate (SDS) A polyanionic (negatively charged) detergent containing a hydrophobic group (12-carbon chain) and a hydrophilic group (sulfate); commonly used to denature proteins.

specific activity With regard to radioactive molecules, the ratio of radioactivity to mass; with regard to enzymes, the ratio of enzyme activity in a sample to the total mass of some component (such as protein or DNA) in that sample.

S-phase The period of a eukaryotic cell cycle during which the DNA is replicated.

substrate A molecule that is modified by an enzyme in a reversible chemical reaction.

S-value   Sedimentation coefficient, usually ascribed to macromolecules and subcellular particles; the S-value is proportional directly to a particle's mass (m) and indirectly to its frictional coefficient (f) according to the equation $S = km/f$.

superhelical turns   Turns of the double-helical DNA molecule introduced before the covalent closure of the two ends of the DNA.

supernatant (sup, supernate)   The solution obtained after formation of the pellet during centrifugation; generally obtained by decantation or aspiration.

T cells   White blood cells that originate in the bone marrow, mature in the thymus gland, and play a major role in the immune defense against cells such as fungi and bacteria. *See* **lymphocyte**.

temperature sensitive (ts) mutant   A cell or virus that possesses a mutation such that it exhibits a particular function at a certain low temperature, but not at a somewhat higher temperature.

terminal differentiation   The final state of cell differentiation, usually with concomitant loss in ability to replicate (divide).

terminator sequence   A nucleotide sequence in DNA that causes RNA polymerase to cease transcription.

termini, 5′ and 3′   The 5′ terminus of DNA or RNA is the end with a free hydroxyl or phosphate group attached to the number 5 carbon of the terminal sugar;

the other end is called the 3′ terminus because the free hydroxyl or phosphate group is attached to the number 3 carbon of the sugar.

tetraploid   Having four times the number of chromosomes as that found in the corresponding haploid.

thalassemia   A very rare family of anemic disorders, all characterized by a reduced rate of synthesis of one or more of the various types of protein chains (globins) found in hemoglobin.

thymus   A ductless endocrine gland in the neck region of mammals that is involved in the maturation of T cells.

titer   A concentration of an infectious agent, such as a virus, or of an antibody.

transcription   The copying of genetic information encoded in the nucleotide sequence of DNA into a nucleotide sequence of an RNA molecule by RNA polymerase.

transcription unit   A section of DNA that is transcribed into a single RNA molecule. A promoter sequence is near the beginning and a terminator sequence is at the end of the DNA in a transcription unit.

transfection   The incorporation of genetic material (DNA) from a donor organism into a recipient cell, which can be of a different species from the donor.

transformed cell   A cell that in culture possesses properties that are characteristic of tumorigenic cells in a cancer.

transgenic mouse   A mouse that carries heritably transmissible foreign genes; that is, genes that have been introduced, usually from some other species, into a fertilized mouse egg cell.

translation The transfer of information from an RNA molecule into a polypeptide, involving a change of "language" from nucleic acids to amino acids; occurs on ribosomes.

trisomic Pertaining to a cell, tissue, or organism having one chromosome represented three times; for example, trisomy 21 implies that chromosome number 21 is represented in triplicate.

tumorigenic cell A cell with the genetic potential for causing cancer.

vector An agent that transfers genes; plasmid vectors transport recombinant DNA from one type of cell to another, and insect vectors transmit viruses from plant to plant.

vesicle A membrane-enclosed structure that is generally formed by the pinching off of membranes from various cellular organelles.

virion A complete virus particle.

Credits

Chapter 1 Figures. 1.1: Campbell, N. (1987). Biology. Menlo Park, Calif.: Benjamin/Cummings, p. 121. 1.2: Starr, C., and R. Taggart (1984). *Biology* (3rd ed.), Belmont, Calif.: Wadsworth, p. 63. 1.3, 1.4, 1.5: Bessis, M. (1973). *Living Blood Cells and Their Ultrastructure*. New York: Springer-Verlag, pp. 106, 242, 244. 1.6: Ambros, V., L. B. Chen, and J. M. Buchanan (1975). Surface ruffles as markers for studies of cell transformation by Rous sarcoma virus. *Proceedings of the National Academy of Sciences USA*, 72:3144. 1.7: Ash, J. S., P. K, Vogt, and S. J. Singer (1976). Reversion from transformed to normal phenotype by inhibition of protein synthesis in rat kidney cells infected with a temperature-sensitive mutant of Rous sarcoma virus. *Proceedings of the National Academy of Sciences USA*, 73:3603. 1.8: Morgan, C., and H. M. Rose (1986). Structure and development of viruses as observed in the electron microscope. *Journal of Virology*, 2:925. 1.9: Koenig, S., H. E. Gendelman, J. M. Orenstein, M. C. Dal Canto, G. H. Pezeshkpour, M. Yungbluth, F. Janotta, A. Aksamit, M. A. Martin, and A. S. Fauci (1986). Detection of AIDS virus in macrophages in brain tissue from AIDS patients with encephalopathy. *Science*, 233:1089. 1.10: Wolfe, S. L. (1985). *Cell Ultrastructure*. Belmont, Calif.: Wadsworth, p. 10. 1.11: Darnell, J., H. Lodish, and D. Baltimore (1986). *Molecular Cell Biology*. New York: Scientific American Books, p. 575. 1.12: Starr, C., and R. Taggart (1987). *Biology*. Belmont, Calif.: Wadsworth, p. 90; electron micrograph inset courtesy of D. Branton. 1.13B: Frye, L. D., and M. Edidin (1970). The rapid intermixing of cell surface antigens after formation of mouse-human heterokaryons. *Journal of Cell Science*, 7:319. 1.14: Darnell, J., H. Lodish, and D. Baltimore (1986). *Molecular Cell Biology*. New York: Scientific American Books, p. 591. 1.15, 1.16, 1.17: Schlessinger, J., Y. Shechter, M. C. Willingham, and I. Pastan (1978). Direct visualization of binding, aggregation, and internalization of insulin and epidermal growth factor on living fibroblastic cells. *Proceedings of the National Academy of Sciences USA*, 75:2659. 1.18: Kleinschmidth, A. K., D. Lang, D. Jacherts, and R. K. Zahn (1962). Preparation and length measurements of the total deoxyribonucleic acid content of T₂ bacteriophages. *Biochim. Biophys. Acta*, 61:857. 1.19: Cairns, J. (1963). The chromosome of *Escherichia coli. Cold Spring Harbor Symposium on Quantitative Biology*, 28:43. 1.20: Kavenoff, R., L. C. Klotz, and B. H. Zimm (1973). On the nature of chromosome-sized DNA molecules. *Cold Spring Harbor Symposium on Quantitative Biology*, 38:1. 1.21, 1.22: Hourcade, D., D. Dressler, and J. Wolfson (1973). The nucleolus and the rolling circle. *Cold Spring Harbor Symposium on Quantitative Biology*, 38:537. 1.23: Salzman, N. P., G. C. Fareed, E. D. Sebring, and M. M. Thoren (1973). The mechanism of SV40 DNA replication. *Cold Spring Harbor Symposium on Quantitative Biology, 38*:257. 1.24: Griffith, J. D. (1975). Chromatin structure: deduced from a minichromosome. *Science, 187*:1202. 1.25: Scheer, U., and H. Zentgraf (1978). Nucleosomal and supranucleosomal organization of transcriptionally inactive rDNA circles in *Dytiscus* oocytes. *Chromosoma, 69*:243. Also Gall, J. G. (1981). Chromosome structure and the C-value paradox. *Journal of Cell Biology, 91*:3S. 1.26: Campbell, N. (1987). *Biology.* Menlo Park, Calif.: Benjamin/Cummings, p. 378. Electron micrographs: (A) From Fig. 1.25; (B) and (C) courtesy of A. L. Olins and D. E. Olins; (D) and (E) courtesy of D. A. Agard; (F) courtesy of B. A. Hamkalo. 1.27, 1.28, 1.29: Darnell, J., H. Lodish, and D. Baltimore (1986). *Molecular Cell Biology.* New York: Scientific American Books, pp. 375, 377, 382.

Chapter 2   Figures. 2.1: Darnell, J., H. Lodish, and D. Baltimore (1986). *Molecular Cell Biology.* New York: Scientific American Books, p. 159. Also Alberts, B., D. Bray, J. Lewis, M. Raff, K. Roberts, and J. D. Watson (1983). *Molecular Biology of the Cell.* New York: Garland, p. 167. 2.6: Photograph courtesy of Beckman Corp. 2.7: Farron-Furstenthal, F. (1975). Protein kinases in hepatoma, and adult and fetal liver of the rat. I. Subcellular distribution. *Biochemical and Biophysical Research Communications, 67*:307. 2.9: Lehninger, A. (1975). *Biochemistry* (2nd ed.). New York: Worth, p. 162. 2.10: Darnell, J., H. Lodish, and D. Baltimore (1986). *Molecular Cell Biology.* New York: Scientific American Books, p. 233. 2.11: Karp, G (1979). *Cell Biology.* New York: McGraw-Hill, p. 114. 2.12, 2.13: Walsh, D. A., J. P. Perkins, and E. G. Krebs (1968). An adenosine 3′,5′-monophosphate-dependent protein kinase from rabbit skeletal muscle. *Journal of Biological Chemistry, 243*:3763. 2.14: Lehninger, A. (1975). *Biochemistry* (2nd ed.) New York: Worth, p. 814. 2.16: Collett, M. S., and R. L. Erikson (1978). Protein kinase activity associated with the avian sarcoma virus *src* gene product. *Proceedings of the National Academy of Sciences USA, 75*:2021. Tables. 2.1: Farron-Furstenthal, F. (1975). Protein kinases in hepatoma, and adult and fetal liver of the rat. I. Subcellular distribution. *Biochemical and Biophysical Research Communications, 67*:307. 2.2: Farron-Furstenthal, F., and J. R. Lightholder (1977). The purification of nuclear protein kinase by affinity chromatography. *FEBS Letters, 84*:313. 2.3: Collett, M. S., and R. L. Erickson (1978). Protein kinase activity associated with the avian sarcoma virus *src* gene product. *Proceedings of the National Academy of Sciences USA, 75*:2021.

Chapter 3   Figures. 3.2, 3.3: Stent, G. S. (1963). *Molecular Biology of Bacterial Viruses.* San Francisco: W. H. Freeman and Co., pp. 106, 134, 135, 3.4: Darnell, J., H. Lodish, and D. Baltimore (1986). *Molecular Cell Biology.* New York: Scientific American Books, p. 230. 3.5B, 3.6, 3.7: Brenner, S., F. Jacob, and M. Meselson (1961). An unstable intermediate carrying information from genes to ribosomes for protein synthesis. *Nature, 190*:576. 3.8, 3.9: Hall, B. D., and S. Spiegelman (1961). Sequence complementarity of T2-DNA and T2-specific RNA. *Proceedings of the National Academy of Sciences USA, 47*:137. 3.10: Darnell, J., H. Lodish, and D. Baltimore (1986). *Molecular Cell Biology.* New York: Scientific American Books, p. 230. 3.11: Sagik, B. P., M. H. Green, M. Hayashi, and S. Spiegelman (1962). Size distribution of "informational" RNA. *Biophysical Journal, 2*:409. 3.12: Astrachan, L., and E. Solvin (1959). Effects of chloramphenicol on ribonucleic acid metabolism in T2-infected *Escherichia coli. Biophysical Acta, 32*:449. 3.13: Perry, R. P., E. Bard, B. D. Hames, D. E. Kelley, and U. Schibler (1976). The relationship between hnRNA and mRNA. *Progress in Nucleic Acid Research in Molecular Biology, 19*:275. 3.14, 3.15, 3.16: Warner, J. R., P. M. Knopf, and A. Rich (1963). A multiple ribosomal structure in protein synthesis. *Proceedings of the National Academy of Sciences USA, 49*:122.

3.17, 3.18: Tilghman, S. M., P. J. Curtis, D. C. Tiemeier, P. Leder, and C. Weissmann (1978). The intervening sequence of a mouse β-globin gene is transcribed within the 15S β-globin mRNA precursor. *Proceedings of the National Academy of Sciences USA*, 75:1309. 3.19: Miller, O. L., Jr. (1981). The nucleolus, chromosomes, and visualization of genetic activity. *Journal of Cell Biology*, 91:15S. Also Miller, O. L., Jr., and B. R. Beatty (1969). Visualization of nucleolar genes. *Science*, 164:955. 3.20: Foe, E. E., L. E. Wilkinson, and C. D. Laird (1976). Comparative organization of active transcription units in *Oncopheltus fasciatus*. *Cell*, 9:131.

Chapter 4 Figures. 4.1: Manthorpe, M., S. Skaper, and S. Varon (1980). Purification of mouse Schwann cells using neurite-induced proliferation in serum-free monolayer culture. *Brain Research*, 196:467. 4.2: Varon, S., M. Manthorpe, L. R. Williams, and F. H. Gage (1988). Neurotrophic factors and their involvement in the adult CNS. In *Aging and the Brain* (R. Terry, ed.), p. 259. New York: Raven Press. 4.3: Richler, C., and D. Yaffe (1970). The *in vitro* cultivation and differentiation capacities of myogenic cell lines. *Developmental Biology*, 23:1. 4.4, 4.5: Shainberg, A., G. Yagil, and D. Yaffe (1971). Alterations of enzymatic activities during muscle differentiation *in vitro*. *Developmental Biology*, 25:1. 4.6: Fawcett, D. W. (1981). *The Cell* (2nd ed.). Philadelphia: W. B. Saunders, pp. 72, 73. 4.8: Bender, M. A., and D. M. Prescott (1962). DNA synthesis and mitosis in cultures of human peripheral leukocytes. *Experimental Cell Research*, 27:221. 4.9: Hayflick, L. (1965). The limited *in vitro* lifetime of human diploid cell strains. *Experimental Cell Research*, 37:614. 4.10: Martin, G. M., C. A. Sprague, and C. J. Epstein (1970). Replicative life-span of cultivated human cells: effects of donor's age, tissue, and genotype. *Laboratory Investigations*, 23:86. 4.11: Temin, H., and H. Rubin (1958). Characteristics of an assay for Rous sarcoma virus and Rous sarcoma cells in tissue culture. *Virology*, 6:669. 4.12: Macpherson, I., and L. Montagnier (1964). Agar suspension culture for the selective assay of cells transformed by polyoma virus. Virology, 23:291. 4.13: Sutherland, B. M., J. S. Cimino, N. Delihas, A. G. Shih, and R. P. Oliver (1980). Ultraviolet light-induced transformation of human cells to anchorage-independent growth. *Cancer Research*, 40:1934. 4.14: Burkitt, D. P., and D. H. Wright (1970). *Burkitt's Lymphoma*. Edinburgh, Scotland: E. and S. Livingstone, p. 11. 4.15: Epstein, M. A., G. Henle, B. G. Achong, and Y. M. Barr (1965). Morphological and biological studies on a virus in cultured lymphoblasts from Burkitt's lymphoma. *Journal of Experimental Medicine*, 121:761. 4.16: Stiles, C. D., W. Desmond, Jr., G. Sato, and M. H. Saier (1975). Failure of human cells transformed by simian virus 40 to form tumors in athymic nude mice. *Proceedings of the National Academy of Sciences USA*, 72:4971. 4.17, 4.18: Stanbridge, E. J., R. R. Flandermeyer, D. W. Daniels, and W. A. Nelson-Rees (1981). Specific chromosome loss associated with the expression of tumorigenicity in human cell hybrids. *Somatic Cell Genetics*, 7:699. 4.19, 4.20: Vogt, M., J. Lesley, J. Bogenberger, S. Volkman, and M. Haas (1986). Coinfection with viruses carrying the v-Ha-*ras* and v-*myc* oncogenes leads to growth factor independence by an indirect mechanism. *Molecular Cell Biology*, 6:3545.

Chapter 5 Figures. 5.1, 5.5: Campbell, N. (1987). *Biology*. Menlo Park, Calif.: Benjamin/Cummings, pp. 397, 401. 5.2, 5.4: Darnell, J., H. Lodish, and D. Baltimore (1986). *Molecular Cell Biology*. New York: Scientific American Books, pp. 247, 249. 5.6: Gueriguian, J. L. (ed.) (1981). *Insulins, Growth Hormone, and Recombinant DNA Technology*. New York: Raven Press, p. 19. 5.7: Macleod, A., and K. Sikora (eds.) (1984). *Molecular Biology and Human Disease*. Oxford, England: Blackwell Scientific Publications, p. 244. 5.9, 5.11: Prentis, S. (1984). *Biotechnology: A New Industrial Revolution*. New York: George Braziller, pp. 46, 51. 5.10: Cherfas, J. (1982).

Man-Made Life. New York: Pantheon Books, p. 102, 5.13: Gueriguian, J. L. (ed.) (1981). *Insulins, Growth Hormone, and Recombinant DNA Technology.* New York: Raven Press, p. 77. 5.14: Cherfas, J. (1982). *Man-Made Life.* New York: Pantheon Books, p. 159. Also Ullrich, A., J. Shine, J. Chirgwin, R. Pictet, E. Tischer, W. J. Rutter, and H. M. Goodman (1977). Rat insulin genes: construction of plasmids containing the coding sequences. *Science, 196:*1313. 5.15: Cherfas, J. (1982). *Man-Made Life.* New York: Pantheon Books, p. 162. Also Villa-Komaroff, L., A. Estratiadis, S. Broome, P. Lomedico, R. Tizard, S. P. Naber, W. L. Chick, and W. Gilbert (1978). A bacterial clone synthesizing proinsulin. *Proceedings of the National Academy of Sciences USA, 75:*3727. 5.16: Cherfas, J. (1982). *Man-Made Life.* New York: Pantheon Books, p. 164. Also Goeddel, D. V., D. G. Kleid, F. Boliver, H. L. Heyneker, D. G. Yansura, R. Crea, T. Hirose, A. Kraszewski, K. Itakura, and A. D. Riggs (1979). Expression in *Escherichia coli* of chemically synthesized genes for human insulin. *Proceedings of the National Academy of Sciences USA, 76:*106. 5.18: Cherfas, J. (1982). *Man-Made Life.* New York: Pantheon Books, p. 166. Also Williams, D. C., R. M. Van Frank, W. L. Muth, and J. P. Burnett (1982). Cytoplasmic inclusion bodies in *Escherichia coli* producing biosynthetic human insulin proteins. *Science, 215:*687. 5.19, 5.20: Macleod, A., and K. Sikora (eds.) (1984). *Molecular Biology and Human Disease.* Oxford, England: Blackwell Scientific Publications, pp. 138, 140. 5.21, 5.22: Camper, S. A. (1987). Research applications of transgenic mice. *BioTechniques, 5:*638. Table. 5.1: White, R., and J.-M. Lalouel (1988). Chromosome mapping with DNA markers. *Scientific American, 258:*40.

Index

Page numbers in *italics* indicate illustrations.

DATE DUE	
DEC 15 1997	

GAYLORD · PRINTED IN U.S.A.